48th Conference on the International Federation of Knitting Technologies (IFKT 2016)

IOP Conference Series: Materials Science and Engineering Volume 141

Moenchengladbach, Germany
8 – 11 June 2016

Editors:

Susanne Aumann
Andrea Ehermann
Marcus O. Weber

ISBN: 978-1-5108-2820-9
ISSN: 1757-8981

Printed from e-media with permission by:

Curran Associates, Inc.
57 Morehouse Lane
Red Hook, NY 12571

Printed by Curran Associates, Inc. (2016)

For permission requests, please contact the Institute of Physics
at the address below.

Institute of Physics
Dirac House, Temple Back
Bristol BS1 6BE UK

Phone: 44 1 17 929 7481
Fax: 44 1 17 920 0979

techtracking@iop.org

Additional copies of this publication are available from:

Curran Associates, Inc.
57 Morehouse Lane
Red Hook, NY 12571 USA
Phone: 845-758-0400
Fax: 845-758-2633
Email: curran@proceedings.com
Web: www.proceedings.com

TABLE OF CONTENTS

IOP Conf. Series: Materials Science and Engineering **141** (2016) 011001 doi:10.1088/1757-899X/141/1/011001

Editors:

Susanne Aumann, Niederrhein University of Applied Sciences, Germany

Andrea Ehrmann, Bielefeld University of Applied Sciences, Germany

Marcus O. Weber, Niederrhein University of Applied Sciences, Germany

IOP Conf. Series: Materials Science and Engineering **141** (2016) 011001 doi:10.1088/1757-899X/141/1/011001

Preface

During June 09-10th 2016 the "48th International Congress of the International Federation of Knitting Technologists (IFKT)" coincided with IFKTs' 60th anniversary. The Congress was held together with the national conference of the Research Institute for Textile and Clothing from the Faculty of Textile and Clothing Technology, Mönchengladbach "MG Open Spaces" and the "Members' Conference of the Verein Deutscher Textilveredlungsfachleute (VDTF)" (an association of experts in the field of textile finishing). The whole event lasted from 09 until 11th June 2016 and took place in Mönchengladbach and Cologne, Germany. More than 220 people from universities, companies and other fields out of 15 different nations participated in the congress

The congress program consisted of seven keynote speakers, 45 oral presentations in 7 sections, 11 open spaces workshops in English and German, including a special workshop at Monforts' Advanced Technology Center (ATC) on "Knitting Concepts". Furthermore, a design showroom and a poster session were part of the program.

This issue of the IOP Conference Series, Materials Science and Engineering (MSE), records papers of the Congress selected by the Congress Scientific Committee. Additional and further papers, information, articles as well as presentations were submitted to the participants after the Congress by download option.

Without the tremendous efforts and dedication of many people from all parties involved in both the planning and organization, the joint congress would have been impossible. We express also our thanks to the impressive team of reviewers, authors and presenters for their contributions to the success of this event. We also thank our sponsors very much for their generous support.

We invite all participants to the next Congresses of IFKT (please see www.knittingfed.com).

Prof. Dr. Marcus O. Weber
Niederrhein University, Mönchengladbach, Germany
Secretary General of IFKT

IOP Conf. Series: Materials Science and Engineering **141** (2016) 011002 doi:10.1088/1757-899X/141/1/011002

Peer review statement

All papers published in this volume of *IOP Conference Series: Materials Science and Engineering* have been peer reviewed through processes administered by the proceedings Editors. Reviews were conducted by expert referees to the professional and scientific standards expected of a proceedings journal published by IOP Publishing.

Energy harvesting "3-D knitted spacer" based piezoelectric textiles

S Anand, N Soin, T H Shah and E Siores

Institute for Materials Research and Innovation, University of Bolton, Bolton, UK

E-mail: Sca1@bolton.ac.uk

Abstract. The piezoelectric effect in Poly(vinylidene fluoride), PVDF, was discovered over four decades ago and since then, significant work has been carried out aiming at the production of high β-phase fibres and their integration into fabric structures for energy harvesting. However, little work has been done in the area of production of "true piezoelectric fabric structures" based on flexible polymeric materials such as PVDF. In this work, we demonstrate "3-D knitted spacer" technology based all-fibre piezoelectric fabrics as power generators and energy harvesters. The knitted single-structure piezoelectric generator consists of high β-phase (~80%) piezoelectric PVDF monofilaments as the spacer yarn interconnected between silver (Ag) coated polyamide multifilament yarn layers acting as the top and bottom electrodes. The novel and unique textile structure provides an output power density in the range of 1.10-5.10 μWcm^{-2} at applied impact pressures in the range of 0.02-0.10 MPa, thus providing significantly higher power outputs and efficiencies over the existing 2-D woven and nonwoven piezoelectric structures. The high energy efficiency, mechanical durability and comfort of the soft, flexible and all-fibre based power generator is highly attractive for a variety of potential applications such as wearable electronic systems and energy harvesters charged from ambient environment or by human movement.

Keywords: Poly(vinylidene fluoride) PVDF, β phase, energy harvesting, piezoelectric effect

1. Introduction

The harvesting of waste energy from ambient environment and human movement has long been considered as an attractive alternative over traditional rechargeable batteries for providing electrical power to low-energy consumption devices such as wireless body worn sensors and wearable consumer electronics [1-3]. Recently, inorganic nanowires of ZnO, InN, GaN, CdS, ZnS and PZT have shown remarkable ability to harvest energy from small mechanical movements and have shown higher energy conversion efficiency as compared to their micro and macro sized counterparts which was attributed to size-effects, decreased defects and improved mechanical flexibility [1-3]. However, it should be noted that these materials are quite brittle in nature, work only at small levels of strain (< 1 %) and are much harder to integrate on a large scale and quite expensive to produce [1-3]. Moreover, for wearable applications, the energy harvesting devices should provide the right "feel" and comfort as well to the wearer. The current problems associated with flexible piezoelectric generators are; (i) low throughput and long and tedious processing techniques; (ii) low output power densities; and (iii) lack of integration techniques [1-3]. Moreover, for achieving truly integrated piezoelectric materials in textile structures, the durability and the "feel" of the piezoelectric structures needs to be tailored. For energy harvesting from human movement, the fibre based electrical power generators are highly desirable as they are light weight and comfortable and look no different from the conventional fabrics. The conjunction of piezoelectric materials in fibres and therefore fabrics offers a simple route for the building of soft piezoelectric generators. The flexible textile structures can themselves be designed so as to provide piezoelectric output on low levels of strains and loadings while providing high fatigue resistance under a large number of variable mechanical deformation and loading cycles. To address this, we have developed three-dimensional piezoelectric fabrics based on "3-D knitted spacer" textile technology [4].

2. Experimental

2.1. Continuous melt-spinning extrusion of poled PVDF monofilaments

Melt spinning of PVDF homopolymer SOLEF 1008, supplied by Solvay Solexis Ltd., was used for the processing of PVDF fibres on a bi-component pilot plant extruder. The polymer pellets are transported across the length of the heated barrel (temp. profile: 190, 200, 210, 220 °C, across the four heated barrel zones) via a screw, where the melt exits the monofilament spinneret ($\Phi = 0.8$ mm, 230 °C) at a pressure of approx. 100-105 bar where the filament is cooled down by using an air quench operating at 20 °C. At this point, the melt is taken up by a draw down godet rotating at 101 mpm (metres per minute). The filament is then passed upon two pairs of heated godets, rotating at 102 and 505 mpm, respectively, producing the change in the orientation of the chains and the crystalline structure of the filament. The draw ratio of 5:1 was found to be the optimal value of the drawing where a high β phase (~ 80 %) was observed and the filament could be processed continuously. While the filaments are being drawn between 2nd and 3rd pair of godets, an electric field of 0.6 MV/m is applied across the filament between two electrodes to further enhance the β phase of the filaments [4-7].

2.2. Knitting of 3-D spacer piezoelectric fabrics

The fabrics were knitted on an E20 circular weft knitting double-jersey spacer machine with a 30ʺ dia., at a machine speed of 30 rpm (at Baltex Speciality Knitters Ltd U.K.). Figure 1(a, b) illustrates the specific knitted structure produced in this work by using three different yarns; (i) conductive yarn A; (ii) insulating yarn B; and (iii) piezoelectric yarn C. The conductive yarn A (Shieldex® Ag coated PA66, 143/34 dtex, resistivity of <1kΩ/m, supplied by Statex GmbH), is plated on the outside of each of the fabric face; the insulating yarn B (84 dtex, false-twist texturized polyester yarn) is plated inside the structure in such a manner that it shows on the inside of the two fabric faces and finally the piezoelectric monofilament spacer yarn C (300 dtex), is tucked inside the two fabric faces [4, 5]. It should be noted that yarn C is tucked on both cylinder and dial needles in such a manner that it does not protrude through either of the fabric faces and always remains inside the two faces keeping them apart. The total thickness of the fabric structures developed in this work was approximately 3.5mm; however, it should be noted that this thickness can be varied between 2mm and 60 mm, depending on the type of knitting machine (warp or weft knitting) used and the end-use application.

3. Results and discussion

For the all fibre piezoelectric generator produced in this work, with an effective area of 15 cm x 5.3 cm, the peak values of the open-circuit voltage and current were found to be 14 V and 29.8 μA, respectively at an applied pressure of 0.106 MPa. In fact, the total power output increases from 0.08 mW to 0.4 mW over the measured impact range of 0.02 MPa to 0.106 MPa, as shown in Figure 2(a) to (c).

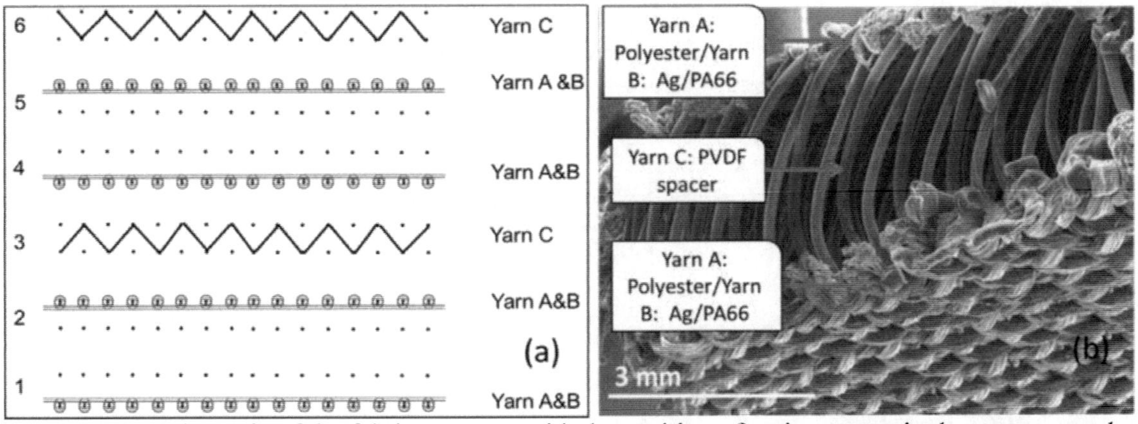

Figure 1. (a) Schematic of the fabric structure with the position of various yarns in the structure and (b) cross-sectional SEM image of the actual fabric clearly showing the position of piezoelectric and conductive yarns [4].

The power output of the piezoelectric fabric can be modelled by using the following equation:

$$Power\ output\ (\mu Wcm^{-2}) = 4.98 - \frac{5.10}{\{1 + exp\ (x - 0.03/0.008)\}}$$

where x is the applied impact pressure and power output is measured in μWcm^{-2}. The 3-D structures provide nearly five times the output power density, with a maximum power density, with a maximum power density of 5.07 μWcm^{-2}, verses 1.18 μWcm^{-2} for knitted 2-D structures, as shown in Figure 2(d).

The power densities obtained from these 3-D piezofabrics are much higher than those reported for 2-D nonwoven $NaNbO_3$–PVDF nanofibre based generators which provide a power output of 2.15 μWcm^{-2}, and electrospun nonwoven PVDF fabrics which generated nearly 3.2 μWcm^{-2} [8]. These values are also significantly higher than those achieved by PVDF nanofibre based generators and $NaNbO_3$ nanowires, which provide an effective power output of 0.115μWcm^{-2} [9]. The excellent performance of our novel generator can be attributed to the following factors; (i) high β phase of the PVDF fibres; (ii) enhanced charge collection due to intimate contact between the PVDF fibres and conductive yarns, leading to improved efficiency; and (iii) transfer of the uniform compression pressure across the fabric surface.

The technologies of knitting of 3-D spacer fabrics have been around for over three decades now and the underlying technology and understanding are highly developed. However, we understand that there have been no previous reports on the use of 3-D spacers fabrics as energy harvesting fabrics and as is the case with every technological development , there are some foreseeable technical challenges which include; (i) optimisation of positioning, spacing and thickness of the spacer piezoelectric yarn and its arrangement in the 3-D structure to enhance the piezoelectric response; (ii) optimization of fabric density and thickness of the spacer fabric for different applications; (iii) ensuring that the conducting yarns used in the opposite faces do not come in contact with each other during the knitting process, or during the cutting procedure ; (iv) as the textiles are intended for use as wearable energy harvesting textiles, the important factors, include air permeability, wicking properties, stretchability and recovery need to be tested and controlled to provide a high level of comfort to the user; and finally (v) the effects of washing, regular use for long periods of time and durability and integrity of the fabric need to be verified to ensure reproducibility of the piezoelectric response and provide a certain lifetime value for the fabric.

IOP Conf. Series: Materials Science and Engineering **141** (2016) 012001 doi:10.1088/1757-899X/141/1/012001

Further work on the coupled mechanical and piezoelectric analysis of the 3-D knitted fabric generators is under way to ascertain the complex relationship between the structure of the fabric and power output and to enhance it further.

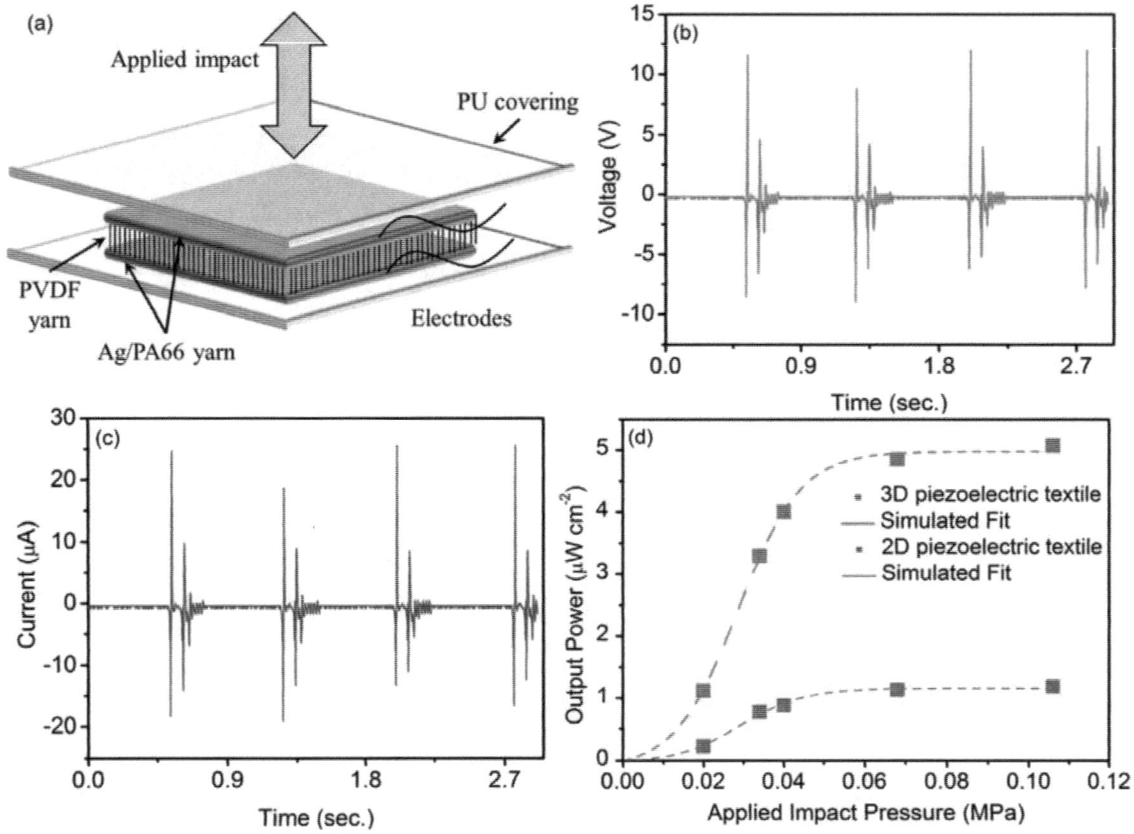

Figure 2. (a) Schematic structure of the packaged 3-D piezoelectric fabric power generator; typical (b) voltage and (c) current outputs of the 3-D piezoelectric fabric (obtained at an impact pressure of 0.034 MPa across a 470 kΩ load); and (d) variation of total output power as a function of applied impact pressure for 2-D and 3-D piezoelectric fabrics [4].

4. Conclusions

A novel knitted all fibre piezoelectric nanogenerator comprising of principally β-phase PVDF fibres in a "3-D knitted spacer" structure, with Ag/PA66 yarn as the charge collecting conducting faces, has been designed, developed and fully characterised. The phase change from a predominately α–phase in raw pallets to a nearly 80% β-phase in filaments has been obtained by drawing of the filaments under an applied electric field. The α to β phase transition has been quantified by using a variety of techniques including FTIR, DSC, NMR, XRD and PFM.

The 3-D knitted spacer piezoelectric fabrics exhibit power density in the range of 1.10μWcm^{-2} to 5.10 μWcm^{-2} at applied impact pressures of 0.02 MPa to 0.10 MPa. This all fibre piezoelectric fabric possesses the advantage of efficient charge collection due to intimate contact of electrodes and uniform distribution of pressure on the fabric surface, leading to enhanced performance. Moreover, the "feel" of the all-fibre piezoelectric generator is not very different from any other conventional textile material and is soft and flexible providing potential maximum level of comfort to the wearer. Bearing all these merits in mind, we believe that our method of producing large quantities of high quality piezoelectric yarn and piezoelectric fabric provides an effective option for the development of high performance energy-harvesting textile structures for electronic devices that could be charged from ambient environment or by human movement.

References

[1] Zeng W, Tao X, Chen S, Shang S, Chan HLW, Choy AS 2013 *Energy Environ. Sci.* **6**, 2631
[2] Fang J, Niu H, Wang H, Wang X, Lin T 2013 *Energy Environ. Sci.* **6**, 2196
[3] Lee M, Chen CY, Wang S, Cha SN, Park YJ, Wang ZL 2012 *Adv. Mater.* **24**(13), 1759
[4] Soin N, Shah T, Anand S, Geng J, Pornwannachai W et al. 2014 *Energy Environ. Sci.* **7**(5), 1670-1679
[5] Soin N, Shah TH, Anand SC, Siores E 2013 UK Patent application no. 1313911.8, Univ. of Bolton
[6] Siores E, Hadimani RL and Vatansever D 2012 *WO Patent WO/2012/035350*
[7] Hadimani RL, Bayramol DV, Soin N, Shah T, Qian L et al. 2013 *Smart Mater. Struct.* **22**(7), 075017
[8] Fang J, Niu H, Wang H, Wang X and Lin T 2013 *Energy Environ Sci* **6**, 2196
[9] Fang J, Wang XG and Lin T 2011 *J. Mater. Chem.* **21**, 11088

IOP Conf. Series: Materials Science and Engineering **141** (2016) 012002 doi:10.1088/1757-899X/141/1/012002

Reasons for changes in the value of unit pressure of compression products supporting external treatment

K Kowalski, A Ilska, M Kłonowska

Lodz University of Technology, Zeromskiego Street 116, Lodz, Poland

e-mail: kjkowalski@p.lodz.pl

Abstract. The paper presents the basics of modelling compression products with intended values of unit pressure for body circumferences with fixed and variable radius of curvature. The derived relationships referring to the dimensions of the fabric's circumferences in a relaxed state of the product were based on Laplace law, local values of the radius of curvature, and the characteristics of stretching and relaxing (deformation) of the knitted fabric, described by experimental relation for the stress and relaxation phase for the 6th hysteresis loop, taking into account confidence intervals. The article indicates the possibilities of using 3D scanning techniques of the human body to identify the radius of curvature of various circumference of the human silhouette, for which the intended value of the unit pressure is designed, and quantitative changes in the body deformation due to compression. Classic method of modelling and design of compression products, based on a cylindrical model of the human body does not provide in each case the intended value of unit pressure, according to specific normative requirements, because it neglects the effect of different values of the radius of curvature of the body circumference and the properties of the viscoelastic knitted fabrics. The model and experimental research allowed for a quantitative and qualitative assessment of the reasons for the changes in the value of unit pressure of compression products supporting the process of external treatment.

1. Introduction

Compression therapy is a method of supporting external treatment used among others in case of varicose veins, lymphedema, post-burn and post-surgical scars. An important parameter determining the effectiveness of this method is the value of unit pressure which the product exerts on the user's body. The range of values of this parameter, depending on the type of therapy and the severity of the patient's condition is determined from the medical point of view [1-3]. The unit pressure can be constant along the whole length of the body (e.g. in treating post-burn scars) or graded - as in the prevention of varicose veins and lymphatic edema. Compression products used in adjuvant therapy are designed based on the cylindrical model of the body. Deviations from this model apply in particular to the human trunk, where there are circumferences whose curvature substantially differs from the circle. In most studies [4-7] modelling of unit pressure is based on assuming circular geometry of body circumferences. Ref. [4] presents the results of modelling unit pressure by finite elements method for a cylinder and a cone. However, in [8] the pressure exerted on the legs by the cuffs of socks was evaluated by theoretical and experimental method for the actual geometry of the leg circumference. Research on the influence of radii of curvature body circumference of on the value of the pressure unit is shown in [7].

The aim of this study is to document, on the basis of qualitative and quantitative analyzes, the reasons for changes in the value of unit pressure of compression products resulting from the following factors:

- differences between the cylindrical body model and the actual geometry of the circumference,
- deformability of the body and the resulting changes in body shape under the influence of the compression product,
- characteristics of stress and relaxation in knitted fabrics related to their rheological properties.

2. Basis for modelling compression knitted fabrics

Modelling and designing of compression products is based on Laplace law (1). It describes the relationship between the unit pressure exerted on a cylindrical body model of the circumference G_1 and circumferential force F for a stripe of a knitted fabric of the width s. In order to design the dimensions of a compression product in free state, with the intended value of unit pressure, it is necessary to know the mechanical characteristics of the knitted fabric, in the form of a relation between force and relative elongation $F = f(\varepsilon)$. Relative elongation of the knitted fabric ε on the covered part of the body circumference is described by a well-known dependence (3).

$$P = \frac{2\pi F}{G_1 s} = \frac{F}{R \cdot s} \quad (1)$$

$$F=f(\varepsilon) \quad (2)$$

$$\varepsilon = \frac{G_1 - Go}{Go} \quad (3)$$, where:

P [hPa] - unit pressure,
F [cN] - circumferential force in the stripe of a knitted fabric of width s,
G_1 [cm] - circumferences of the body parts,
s [cm] - width of the fabric stripe,
G_0 [cm] - fabric circumference in free state,
ε - relative elongation of the fabric.

After substituting equations (2) and (3) to equation (1), and after necessary transformation we obtain dependence for the value of the fabric circumference G_o in free state as a function of the required value of pressure P and body circumference G_1 along the length of the covered body part for the given characteristics of the knitted fabric (2).

2.1 Subject of study

Compression products used in the treatment of burns are most often made of plain stitch warp-knitted fabrics with elastomeric threads, whose stitch is shown in Figure 1. The presented fabric is a three-guide knitted fabric composed of a binding stitch made of a textured polyamide silk with a linear density of 78 dtex (76 %), and vertical wefts made of polyurethane yarn with a linear density of 480 dtex (24 %). The parameters describing the fabric are course density $P_r = 720$, wale density $P_k = 154$ and surface mass $G = 244$ g/m^2.

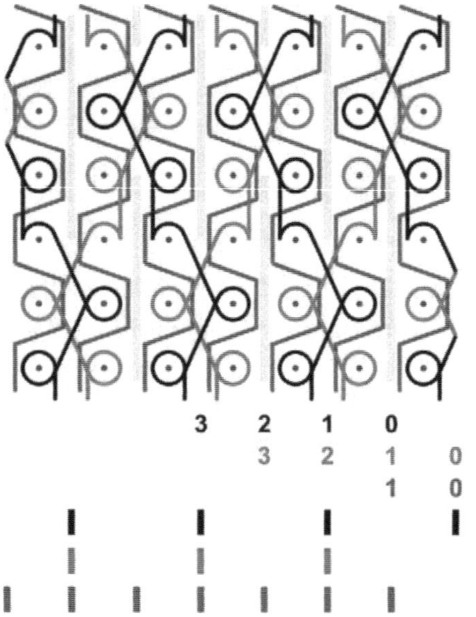

Threading report :

I full set - polyurethane yarn

II 1 in, 1out – polyamide multifilament

III 1 in, 1 out – polyamide multifilament

Recording of model links:

I 11/ 00//
II 34/ 32/34/ 43/ 45/43//
III 32/ 34/ 32/ 23/ 21/ 23

Fig. 1a Record of schematic stitches of highly elastic warp-knitted fabric with elastomeric threads

Fig. 1b Threading report with a record of cell shape

2.2 Mechanical characteristics of the knitted fabric

Research on knitted compression products performed by the procedure shown in [8], was carried out for a range of relative elongation $\varepsilon \in <0; 1>$ in separate stretching areas enlarged by 0.1 of the relative elongation. For each value of the strain, the test was carried out for 5 rectangular samples with a length of 200 mm and width of 75 mm, stitched with a flat seam along the shorter side, cut from different locations of the compression fabric and submitted (by loop method), to stretching and relaxation at a speed of 200 mm/min on the Hounsfield tensile testing machine using needles stabilizing the width of the fabric. For each stretching range 6 hysteresis loops were performed. Figure 2 shows the changes in the average values of the force as a function of the individual ranges of the strain for the 6th hysteresis loop for the stress and relaxation phase, taking into account for, the subsequent relative elongations, maximum forces in the stress phase and minimum forces in the relaxation phase from the confidence intervals. Confidence interval for the mean value is defined by the formula $\bar{x} \pm \frac{U_\alpha S}{\sqrt{n}}$ (5),

wherein: $\bar{x}$ - the arithmetic average counted on the basis of n - elemental sample population, S - experimental estimation of the standard deviation U_α - value of the random variable U of standardized normal distribution (N (0,1)) defined in such a way so as to satisfy the relation (6):

$P\left\{\bar{x} - \frac{u_\alpha \cdot s}{\sqrt{n}} < m < \bar{x} + \frac{u_\alpha \cdot s}{\sqrt{n}}\right\} = 1 - \alpha$ (6), wherein for $1-\alpha = 0.95$; $U_\alpha = 1.96$ [10].

Significant differences between the values of forces in stress and relaxation phase, for the same elongation values result from the rheological properties of the compression fabric. Qualitative interpretations of these differences can be explained on the basis of standard Zener rheological model, according to which the process of forces relaxation is described by the relationship (7).

$F_1 = C \cdot \varepsilon + C_1 \cdot e^{\frac{-t \cdot C_1}{\eta}}$ (7), where:
C, C_1 - relative longitudinal rigidity,
η- relative absolute viscosity

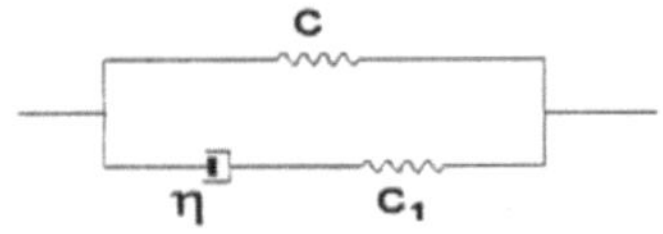

Transferring model interpretations onto the behavior of a compression product during use it should be noted that the circumferential forces in the fabric will tend to be equal to $C\cdot\varepsilon$, because the expression $\exp(-t\cdot C_1/\eta)$ for $t \to \infty$ takes the value 0. This explains the significant differences between the values of forces for the stress and relaxation phases.

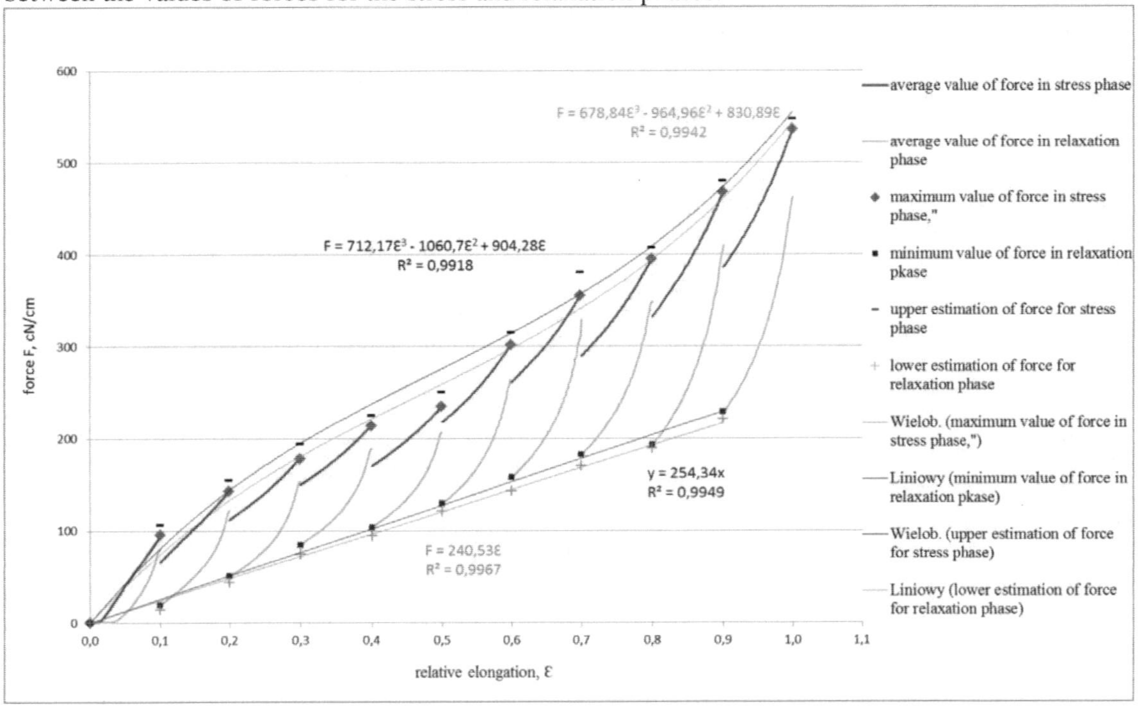

Fig. 2 Characteristics of a knitted fabric – average values of the force as a function of relative elongation for the 6th loop in the stress and relaxation phase with the confidence intervals for the maximum values in the stress phase and minimum values in the relaxation phase

The equations (8.1 ÷ 8.4) referred to the fabric stripe of the width $s = 1$ cm, describe the mechanical characteristics of the fabric in the form of relations between the force F and relative elongation ε for the stress and relaxation phase and from confidence intervals.
Experimental functions for stress phase:

$F = 678,84\cdot\varepsilon^3 - 964,96\cdot\varepsilon^2 + 830,89\cdot\varepsilon$ $R^2 = 0.9942$ for the maximum values (8.1)

$F = 712,17\cdot\varepsilon^3 - 1060,7\cdot\varepsilon^2 + 904,28\cdot\varepsilon$ $R^2 = 0.9918$ for the maximum values from the confidence interval (8.2)

Experimental functions for relaxation phase:

$F = 254.34\cdot\varepsilon$ $R^2 = 0.9949$ for the minimum values (8.3)

$F = 240.53\cdot\varepsilon$ $R^2 = 0.9967$ for the minimum values from the confidence interval (8.4)

3. Influence of the geometry of the circumference on the value of unit pressure

As mentioned above, compression products are designed according to Laplace law. Human body contains numerous circumferences of variable curvatures, and therefore the obtained pressure values are different in places where the radius differs from the one taken into account while designing the product for a cylindrical model of the body part.

Figure 3 depicts the geometry of scanned circumferences selected from the trunk of a female body with and without a compression product. The observed difference in the geometry of the circumferences results from the body's susceptibility to pressure. The measurements were carried out at the Textile Research Institute in Lodz, in Scientific Division Knitting Technologies and Clothing, by the silhouette scanning technique using 3D Body Scanner.

In order to determine the effect of various values of the radius of curvature on the value of unit pressure, the values of R_n radii were determined using the Pythagorean theorem. The radius of curvature was determined for the successive i-points of the circumference as the radius of a circle, approximated on the basis of 5 neighboring points: $(x_{i-2}, y_{i-2}), (x_{i-1}, y_{i-1}), (x_i, y_i), (x_{i+1}, y_{i+1}), (x_{i+2}, y_{i+2})$. Approximation is based on minimizing the mean square error (distance between the points and the circle), while the weight of error at the extreme points is reduced by half. The determined values of the radii of curvature of the given circumference in the range of angles 0 ÷ 360° before and after putting on the compression product, i.e. including the body's susceptibility to pressure are presented in (Figure 4).

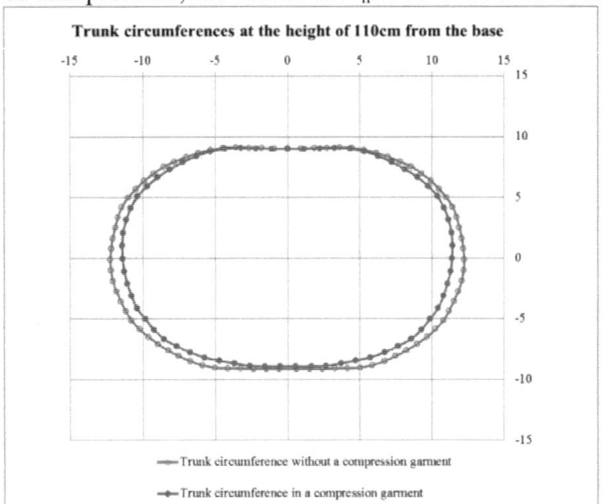

Fig. 3 The geometry of scanned circumferences selected from the trunk of a female body with and without a compression product

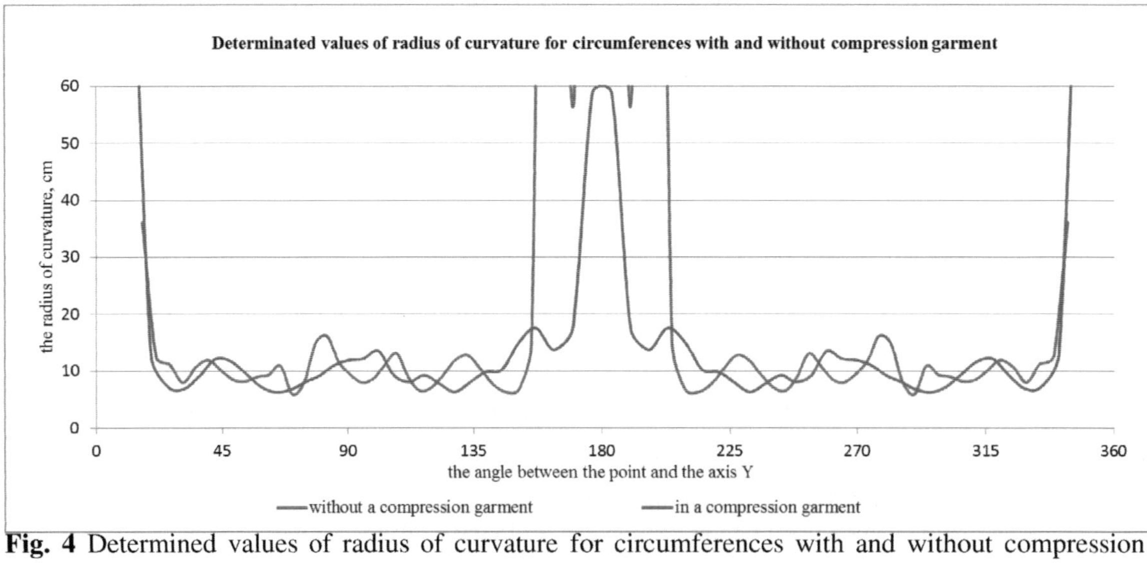

Fig. 4 Determined values of radius of curvature for circumferences with and without compression garment

Figure 5 shows the changes in the unit pressure P for the geometry of the circumference with and without the compression garment. Assuming that the value of the longitudinal force at the circumference is constant, we obtain equation (9) describing the value of unit pressure along the circumference. $P = P_{\text{int}} \cdot \frac{R}{R_n}$ (9), where P_{int} is the intended value of pressure equal to 20hPa, R_n values of the successive radii of curvature of the circumference, R values of radius of curvature of the circumference treated as a circle. As a result of the body's susceptibility to pressure, the circumference with the compression clothing was reduced, resulting in reduction of relative elongation of the fabric and circumferential force. Local changing the radius of curvature of the circumference in compression clothing and the circumferential force causes a change of unit pressure (Figure 5).

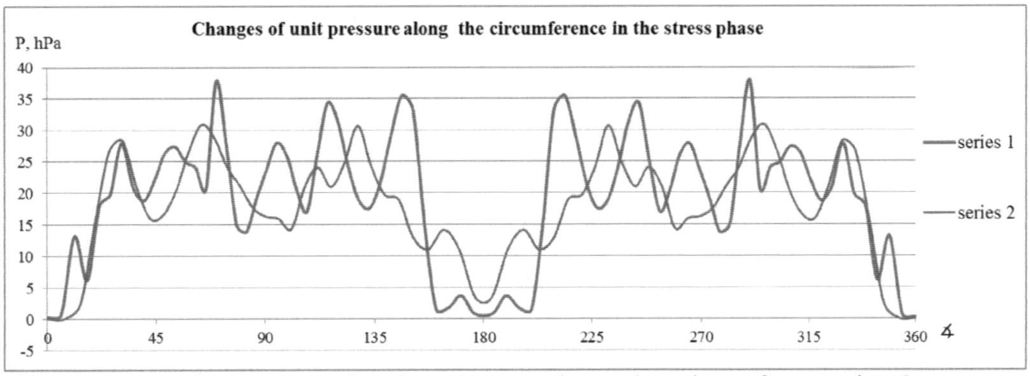

Fig. 5 Examples of changes of unit pressure along the circumference in the stress phase for the following cases:
Series 1 counting parameters: G_1=70.3 cm, R=11.19 cm, P=20 hPa, F = 678.84 ε^3 – 964.96 ε^2 + 830.89 ε
Series 2 counting parameters: G_1=66.8 cm, R=10.63 cm, F = 678,84 ε^3 - 964,96 ε^2 + 830,89 ε

3.1 Percentage changes in the length of the circumferences under the influence of compression product

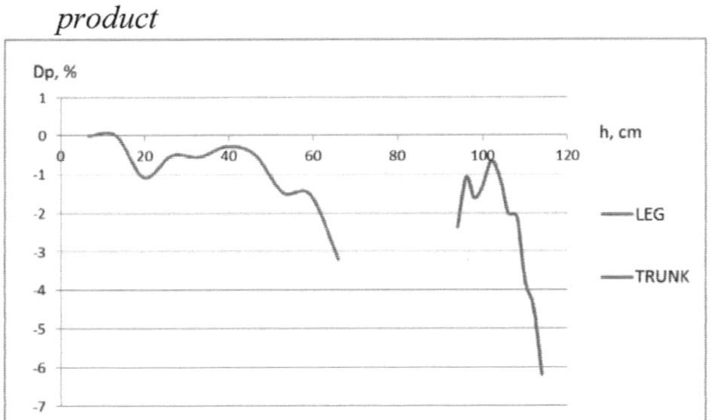

Fig. 6 Percentage changes of legs and trunk circumferences under the influence of a compression product, where:
Dp, % - percentage differences
h, cm - the height from the base

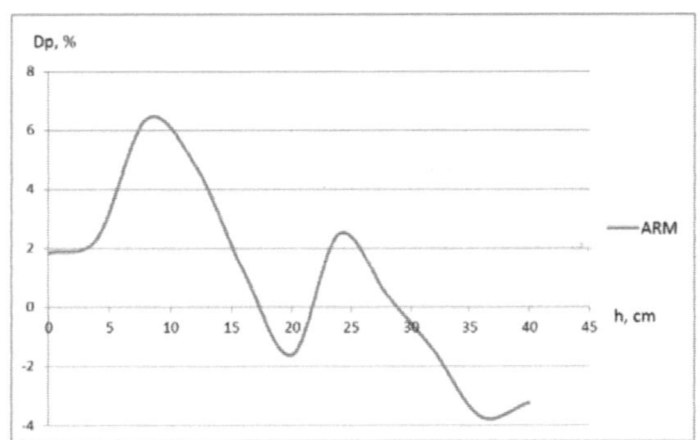

Fig. 7 Percentage changes of hand circumferences under the influence of a compression product, where:
Dp, % - percentage differences
h, cm - the height of the wrist

Figures 6 and 7 show, determined on the basis of the relationship $Dp = \frac{G_2 - G_1}{G_2} \cdot 100, \%$, where:
G_2- circumference in the compression product, cm
G_1- circumference without the compression product, cm
percentage changes in the circumferences of the trunk, legs and arm under the influence of the compression device designed to 20 hPa on the basis of a cylindrical body model.

The biggest changes in the length of circumferences can be observed at the trunk (waist), upper part of the leg (thigh), and the hand. These changes result from the deformability of the body and its deformation under the influence of compression. Depending on the body area and the fat content the percentage changes Dp will differ. Due to the phenomenon of deformation of the geometry of the circumference and the changes in the radius of curvature under unit pressure, parameter Dp should in some cases be taken into account in designing compression products.

4. Changes in the value of unit pressure, taking into account the impact of the force function and relative elongation of the knitted fabric for the stress and relaxation phase

In the next stage of the study the unit pressure changes occurring by the same relative elongation ε for the stress and relaxation phase were determined. In that stage, firstly the values of relative elongation for the body circumference without compression clothing were calculated according to the procedure currently used in the design of compression products of an intended unit pressure P = 20 hPa. Then, for the same value of relative elongation unit pressure values were determined using the functions describing the forces in the relaxation phase $P = \frac{2\pi\cdot(254{,}34\cdot\varepsilon)}{G_1\varepsilon}$ (10).

The results presented in Figure 8 show a significant decline in the value of unit pressure under the influence of stress relaxation. The presented graph shows that only in the stress phase, in those places of the circumference in which the radius of curvature is equal to the radius for which the product was designed, the intended value of the unit pressure was achieved, while in the relaxation phase this value has not been reached.

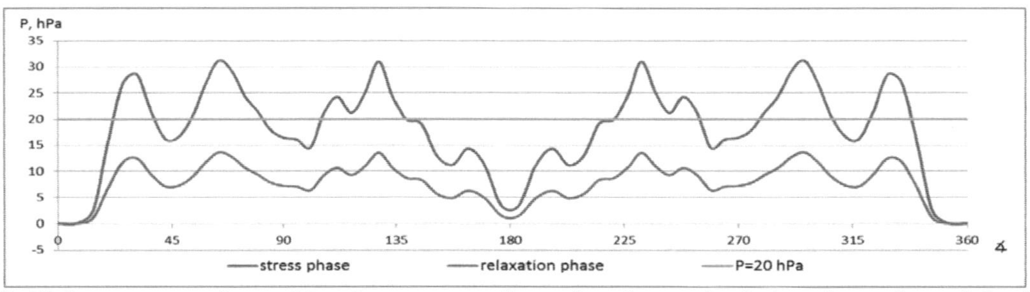

Fig. 8 Unit pressure values taking into account force function and relative elongation of the fabric for stress and relaxation phase. Counting parameters: G_l=66.8cm, R=10.63 cm, ε= 0.33

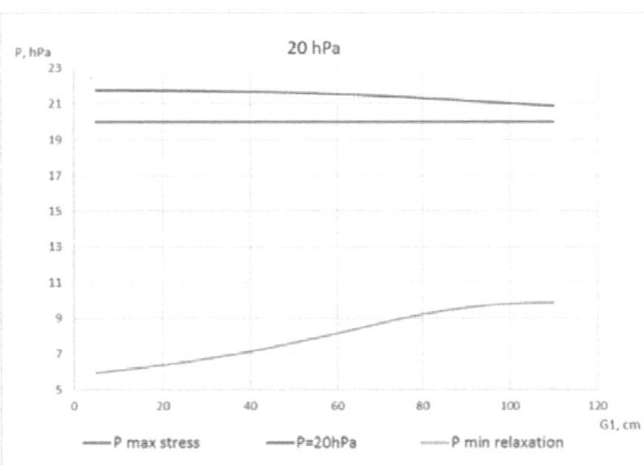

Fig. 9 Changes in the value of unit pressure taking into account the impact of force function and relative elongation of the fabric, using the determined confidence intervals for the stress and relaxation phase as a function of circumference G_l and the intended value 20 hPa.

In the last stage of the research some generalization was made of unit pressure for a wide range of circumferences G_l. In the first stage of calculations, values of relative elongations were determined for the stress phase and the intended pressure P = 20 hPa for each circumference $G_l \in <5;110>$ cm. Then, the calculated values of relative elongation were introduced accordingly to the equations (11), (12) to calculate the value of unit pressure:

- for the stress phase

$$P = \frac{2\pi(712{,}17\cdot\varepsilon^3 - 1060{,}7\cdot\varepsilon^2 + 904{,}28\cdot\varepsilon)}{G_1\varepsilon} \quad (11)$$

- for the relaxation phase

$$P = \frac{2\pi\cdot 240{,}53\cdot\varepsilon}{G_1\varepsilon} \quad (12).$$

The calculated pressure values for the stress and relaxation phase, taking into account the impact of the force function F and relative elongation ε of the knitted fabric, using the determined confidence intervals for both phases as a function of circumference G_l and the intended value 20hPa are presented in Figure 9.

The values of unit pressure in the relaxation phase are significantly lower than the intended ones, which confirms the significant impact of the relaxation phenomenon on the value of unit pressure.

5. Conclusions

On the basis of qualitative and quantitative analyzes the paper presents the changes in the value of unit pressure exerted by compression products on the user's body resulting from the following factors:
a) the characteristics of stress and relaxation of the knitted fabric resulting from its rheological properties. The study shows that the procedure currently valid for determining mechanical characteristics in the form of thc rclation between the force and relative elongation of a knitted fabric limited to the stress phase leads to a significant lowering of values of the unit pressure, because it does not take into account the relaxation processes occurring during the usage of products, which in case of many therapies are worn practically all day and over a period of several months.
b) the actual geometry of body circumferences having different radii of curvature. Design of compression products for circumferences of circular geometry makes it possible to obtain the intended values of unit pressure only in the body areas in which the real radius of curvature is the radius of a circle, which in some cases may reduce the effectiveness of the therapy. Thus, the local value of the radius of curvature should be in certain cases taken into account while designing compression products.
c) the susceptibility of the body to pressure. This factor is the reason for subsequent changes in the values of unit pressure, due to changes in the circumferences and the radii of curvature.

Acknowledgements

This research was financially supported by the Polish National Centre for Research and Development Project No. 244972, Agreement nr PBS3/B9/46/2015

References

[1] Mikołajczyk A, Sośniak K, Fryc D, Miś K 1999 *Strategia postępowania w leczeniu blizn przerostowych u dzieci oparzonych*, Dermatologia Kliniczna i Zabiegowa, 1, supl. 2, 74-76

[2] Fritz K, Gahlen I, Itschert G 1996 *Gesunde Venen – Gesunde Beine,* Rowohlt Taschenbuch Verlag GmbH, Reinbek bei Hamburg

[3] Norms: ISC 11.120.20 prEN 12718 : 1997

[4] Abbas MS, Mansour R, Zeynab S 2008 The Analytical Study of garment Pressure on the Human Body Using Finite Elements, *FIBRES & TEXTILES in Eastern Europe* **3**, 69-73

[5] Maklewska E, Nawrocki A, Ledwoń J, Kowalski K 2006 Modelling and Designing of Knitted Products used in Compressive Therapy, *FIBRES & TEXTILES in Eastern Europe* **14**, 111-113

[6] Nawrocki A, Kowalski K, Maklewska E 2006 *Modelowanie i ocena instrumentalna ucisku wywieranego przez wyroby tekstylne stosowane w leczeniu i rehabilitacji blizn pooparzeniowych,* Research project 4T08 E 05425, Łódź

[7] Dan R, Dan M-H, Fan X-R, Chen D-S, Shi Z, Zhang M 2013 Numerical Simulation of Dynamic Pressure and Displacement for the Top Part of Men's Socks Using the Finite Element Method, *FIBRES & TEXTILES in Eastern Europe* **21**, 112-117

[8] Bland DR 1960 *The Theory of linear Viscoelasticity*, Pergamon Press

[9] Ilska A,Kowalski K, Kłonowska M, Kowalski TM 2014 Influence of Stress and Relaxation Characteristics of Knitted Fabrics on the Unit Pressure of Compression Garments Supporting External Treatment, *Fibres&Textiles in Eastern Europe* **22**, 87-92

[10] Krysicki W i współautorzy, *Rachunek prawdopodobieństwa i statystyka matematyczna w zadaniach,* część I, II, Wydawnictwo Naukowe PWN, Warszawa, 2004

[11] Nawrocki A, Kowalski K, Maklewska E 2000 *Modelling and instrument evaluation of compression exerted by textile garments used in a treatment and rehabilitaion of post-burn scars,* Research project 4T08 E 05425, Lodz

Three-layer knitted materials for protective clothing

E Mielicka, J Janicka, R Kozminska, A Walak
Textile Research Institute, Lodz, Poland

E-mail: emielicka@iw.lodz.pl

Abstract. The results of investigating multifunctional 3D knitted materials dedicated for protective clothing were presented. The 3D design structures were made on a circular knitting machine using yarns with flame retardant or electrostatic properties. The functionality imparted to each of the assortments developed was verified during the tests in accredited laboratories as well as by assessing their biophysical properties. Based on the analysis of the test results, a beneficial effect of the raw materials and the 3D structure of knitted fabrics were demonstrated. Designed garments could be useful as individual protection clothing for workers exposed to harmful occupational environment factors, such as heat and static electricity. The study was conducted within the project EUREKA E! 5799 BATAN "Multifunctional knitted fabrics with barrier properties for clothing".

1. Introduction

The problems of human health and life protection in work environment are a current subject of studies in many research centers. One of them is the development of protective clothing and search for new technological and structural solutions in the processes of their manufacture. Textiles designed for protective clothing gain their desired properties by using proper textile raw materials, designed structures and special finishing treatments.

Among the factors mentioned, a priority issue includes raw materials and their modification to obtain multi-functional features, e.g. high strength and flame-retardant characteristics [1-5]. The analysis of raw material market allows one to select yarns, offered by manufacturers, with special properties, made of the given type of fibers in 100 % as well as in blended systems that owing to their properties can be used in specified applications, such as:

- Protective clothing against fire and high temperatures: Meta-Aramid Nomex® T450, Para-Aramid Kevlar® stretch broken, Panox®, Vectran®, Viscose FR®, Modacryl,
- High resistant clothing protecting against cutting and abrasion: 100% Kevlar®, Kevlar®/glass filament, Kevlar®/steel wire, Vectran®.
- High strength clothing protecting against abrasion and high temperatures: Kevlar®, Vectran®, polyamide, Zylon®,
- Antistatic clothing and filtration materials: 100 % Inox, polyester/Inox, Nomex®/Inox, Kevlar®/Inox, polyamide/Inox, Beltron®, Negastat®.

The use of the yarns mentioned above depends on the basic properties of fibers that are used for their production. Some essential properties for selected types of fibers are presented in Table 1. The characteristics of the given raw materials presented by manufacturers constitute a valuable assistance for the producers of textiles, whose production is connected with fulfilling many specified requirements [6].

Table 1. Selected properties of special fibers used in the production of yarns designed for protective clothing

Parameter / Type of fiber	Most often used fiber linear density [dtex]	Specific gravity [g/cm^3]	Tenacity [cN/tex]	Elongation [%]	Max. temp. that unchanged the fiber structure [°C]	Temperature of essential changes in fiber [°C]
Kevlar®	1.7	1.44	210	3	200	680
Nomex®	2.2	1.38	48	22	200	640
Panox®	2.2	1.36	19	25	200	580
Viscose FR®	2.2	1.52	18	16	170	400
Polyester	2.4	1.38	50	14	130	220
Polyamide 6.6	various	1.14	66	40	110	220
Inox (metal)	7.9	7.9	23	1	690	980
Glass fiber	various	2.6	60-100	2.5	350	820

A significant position within a wide range of protective clothing assortments is occupied by textiles protecting against heat at elevated temperature and static electricity as well as those that meet the requirements of flame-retardant fabrics. An expected feature is also their antibacterial capacity. In this connection, consumers, in relation to the protective clothing offered, point out that their functional properties should comprise multi-functional capabilities including the physiological comport of use. The issue of multi-functionality of fabrics designed for protective clothing has been a subject of many research works at the Textile Research Institute that pays much attention to fabrics made by the technique of knitting [7-9].

In recent years, studies have been carried out on multi-functional knitted fabrics with a multi-layer structure designed for protective clothing or elements of protective, flame-retardant and antistatic clothing. This paper presents selected results of these studies concerning weft-knitted fabrics with a three-layer structure.

2. Technological tests

The aim of the research undertaken was to develop new assortments of knitted fabrics designed for clothing with barrier properties against elevated temperature and static electricity, showing also biological activity concerning antibacterial action. Moreover, the studies were aimed at the multi-functionality of knitted fabrics, including features of improved physiological comfort and an aesthetic appearance expected by users. In order to obtain the multi-functionality of knitted fabrics consisting in imparting to them particular properties, the following tasks were performed:

- Flame-retardant and antistatic properties of fabric were obtained by using for their production yarns of raw materials with the required features.
- To obtain antibacterial features, a chemical agent containing nanoparticles of metallic silver was use in the finishing processes of fabrics.
- To obtain the physiological comfort required with the given barrier properties, there were designed three-layer, weft knitted structures as fabrics with LR stitches (two layers) combined by a middle layer formed of a flame-retardant monofilament yarn. Designing the structures of knitted fabrics, there were used, in particular layers, such types of yarns to obtain the internal layers (in contact with user's skin) of clothing with good properties of physiological comfort [9].

A rib knitting machine, type Ovja from Mayer Company, modernized at the Textile Research Institute, was selected to produce the knitted fabrics to be tested. Its technical parameters were adapted for the production of three-layer knitted fabrics with a set zone of distance. The modernization of the knitting machine constitutes the subject of an improvement application (submitted in 2012) and a part of Patent Application No. P 406 791, 2014, to the Patent Office of the Polish Republic; „Distance 3D weft-knitted fabric and a process and system of its making". The invention consists in that the stitch system designed uses the technique of weft-knitted fabrics of three-layer structures, in which the layers are combined by means of tucking links of monofilament yarn, located between the needles of both beds forming the distance space [10].

In this connection, there were designed three-layer 3D distance knitted fabrics, including a technological program of producing samples of these knitted fabric with the following compositions:

Sample 1. Metaaramid + metaaramid/viscose FR + monofil PES

Sample 2. Metaaramid + viscose FR + monofil PES

Sample 3. Metaaramid + Rezistat* + metaaramid/viscose FR + monofil PES

*/Rezistat – trade name of blended yarn: Vsc Lenzing + PES/carbon compounds

Sample 4. Metaaramid + modakryl Protex/cotton + monofil PES

Sample 5. Metaaramid + viscose FR + monofil PES

Sample 6. Metaaramid /antistatic + metaaramid + wool/viscose FR + monofil PES.

The knitted fabrics were made with the use of combined stitch that resulted in a spatially developed structure owing to the application of the stitch system in the following way: the top layer is formed by means of even RL stitch or RL plated stitch and the under surface by means of RL stitch. Both layers were combined with a monofilament yarn using a half cardigan stitches. The use of monofilament yarn to combine the layers caused the formation of spacing between the fabric layers – a distance of about 3 mm, while the links – yarn sections between the layers formed the middle layer. The preservation of distance in knitted fabrics depends, among others, on the thickness of monofilament yarn and the way of its inserting between the layers. This is essential due to the need for providing fabric elasticity fabrics. In the preparation of the sample, two methods of setting the tucking links were used: in samples 1, 2, 3 and 4, a skew roof setting of the monofilament yarn was used, while in Samples 5 and 6 – an alternating skew setting in particular fabric courses was applied.

Each of the three layers of particular fabric sample was made of different raw material. The top layers were made of aramid (meta-aramid) yarn and antistatic yarn (Rezistat, meta-aramid/antistatic), while the bottom layers – with a predominance of natural and man-made yarns (meta-aramid/viscose FR, viscose FR, modacryl Protex/cotton, wool/viscose FR). In all the samples, the distance layer was made of PES monofilament yarn.

It was assumed that the three-layer knitted fabrics designed in the way mentioned above will show particular biophysical properties creating the effect of micro-climate around the user's body wearing the clothing made their use. In view of expected good thermo-regulating properties, the three-layers knitted fabrics have been planned to use for warming outerwear, such as waistcoats and warming elements in outerwear.

The linear density of the most multi-filament yarns used ranged from 20 to 25 tex, while that of monofilament yarn was 100 dtex. Table 2 presents the percentage content of particular raw materials in the fabric samples made.

The finishing treatment of the three-layer knitted fabrics was carried out according to the technological parameters of finishing knitted fabrics made of synthetic (aramid yarns) and natural and man-made yarns (cotton, wool, viscose yarns) in the following finishing stages: washing, neutralization, rinsing, centrifuging, strip reversing, fabric splitting along needle set-out, drying and stabilization.

- Preliminary stabilization was added to the finishing process as the first phase.
- In the finishing of knitted fabrics no softening agents were used.

In order to impart antibacterial properties to the fabrics, their selected assortments were treated with biologically active agents in the finishing process, using an antibacterial agent Ruco Bac AGP (Sample 4) or an agent containing nanoparticles of metallic silver (Sample 5).

Table 2. Percentage content of particular raw materials in the knitted fabrics

Sample No.	Type of raw material	Content [%]
1.	meta-aramid meta-aramid/viscose FR monofil PES	58.7 2.6 15.7
2.	meta-aramid viscose FR monofil PES	58.7 25.6 15.7
3.	meta aramid Rezistat meta-aramid/viscose FR monofil PES	58.7 5.0 21.0 15.3
4.	meta-aramid Protex/cotton monofil PES	58.4 27.1 14.5
5.	meta-aramid viscose FR monofil PES	50.0 35.0 15.0
6.	meta-aramid/antistatic metaaramid wool/viscose FR monofil PES	26.0 20.0 40.0 14.0

As a result of the assessment of the knitting and finishing processes followed by the organoleptic evaluation of the fabric samples, especially with respect to the fabric spatial structure, Samples 3, 4, 5 and 6 were selected for the tests confirming the expected functional properties.

3. Assessment of the functionality of three-layer knitted fabrics

The assessment of functionality of the knitted fabrics under investigation was carried out with respect to their structural and physico-mechanical properties as well as their physiological comfort and functional barrier features concerning flammability, static electricity and biological activity [9].
The following parameters characterizing the fabric **physical and functional properties** were tested (Table 3):

- Surface mass according to standard *PN–P-04613-1:1997*
- Course and wale stitch density according to standard *PN-EN 14971:2007*
- Changes in dimensions after laundering according to standards *PN - EN ISO 5077:2011, and PN-EN ISO 6330:2002. (A1:2011).*
- Fabric thickness according to standard *PN-EN ISO 5084:1999.*

In compliance with the planned applications of the knitted fabric assortments proposed (outerwear and warming elements in outerwear), in order to assess of their protective properties against hot thermal effects, we used the test methods and requirements given in standard *PN-EN ISO 11612:2011.* This standard determines the following range of tests for the parameters presented below (Tables 4, 5, 6 and 7):

- Resistance to ignition of textiles, the test method according to standard *PN-EN ISO 15025: 2005*
- Resistance to changes under the influence of hot air, the test method according to standard *ISO 17493:2000*
- Determination of the parameter of heat penetration with the action of flame, test method according to *PN-EN 367: 1996*
- Determination of the coefficient of the heat radiation transfer, test method according to standard *PN-EN ISO 6942: 2005.*

For the knitted fabrics, whose raw material system was designed to include antistatic yarns (Samples 3 and 6), **electrostatic properties** were determined, assessing the following parameters (Table 8):

- Surface resistance, test method according to standard *PN-EN 1149-1:2008*
- Volume resistance, test method according to standard *PN-EN 1149-2:1999/Ap1:2001*
- Half decay time and shielding coefficient, test method according to standard *PN-EN 1149-3:2007.*

The assessment on the parameters was performed on the basis of standard *PN-EN 1149-5:2009.*
The **antibacterial properties** of the knitted fabrics (Table 9) after finishing (Sample 4 and 5), with the use of two different antibacterial agents, were assessed by the quantitative method against gram positive bacteria, *Staphylococcus aureus* and gram negative bacteria, *Escherichia coli.* The tests were carried out according to the Testing Procedure of the Textile Research Institute, standards *AATCC Test Method 100-2004* and *PN-EN ISO 20743.*
The **physiological comfort** of the three-layer knitted fabrics was determined with the use of biophysical parameters such as (Table 10):

- Thermal resistance and water vapor resistance according to standard PN-EN 31092:1998/Ap1:2004.
- Air permeability according to standard PN-EN ISO 9237:1998.
- Hygroscopicity according to standard PN-80/P-04635.

Table 3. Physical and functional parameters of three-layer knitted fabrics

Sample No.	Surface mass [g/m^2]	Fabric tightness		Change in dimension after washing		Average thickness [mm]
		Number of courses/cm	Number of wales/cm	lengthwise [%]	crosswise [%]	
Flame-retardant knitted fabrics						
4.	305	12.5	9.0	-1.0	1.0	2.05
5.	376	14.1	9.4	-5.0	-1.0	2.80
Flame-retardant and antistatic knitted fabrics						
3.	294	12.5	9.0	-4.5	-0.5	2.44
6.	434	14.2	9.3	-2.0	-4.0	2.88

Table 4. Test results of the resistance to ignition of textiles

Test results according to procedure A:	
• Does the flame reach the upper or vertical sample edge?	**NO** – all the samples
• Did debris occur?	**NO** – all the sample
• Was debris on fire ?	**NO** – all the samples
• Did a hole occur ?	**NO** – all the samples
• Time of consequent flame combustion [s]	**0 s.** – all the samples
• Time of consequent glow [s]	**0 s.** – all the samples

Assessment: Samples 3, 4, 5, 6 fulfill the requirements of resistance to ignition for range A1.

Table 5. Test results of the resistance of fabric to heat at a temperature of 180 °C

Sample No.	**Change in dimensions**		**Change in appearance**	***Requirements according to PN-EN ISO 11612:2011***
	lengthwise [%]	**Crosswise [%]**		
3.	-1.1	-0.8	No change in appearance	*Sample should not melt, ignite or shrink by more than 5% at a temperature of 180° C ±5.*
4.	-2.4	-4.4	No change in appearance	
5.	-1.6	-3.0	No change in appearance	
6.	-0.5	-2.4	A light yellowing on fabric back	

Assessment: all the fabric samples tested fulfil the required resistance to heat at a temperature of 180°C

Table 6. Test results of the resistance of fabrics exposed to the action of thermal radiation

Sample No.	**Coefficient of thermal radiation transfer RHTI 24**	**Coefficient of thermal heat transfer RHTI 12**	**RHTI 24 - RHTI 12**	***Requirements according to PN-EN ISO 11612:2011***
3.	21.7	13.8	7.9	*RHTI 24* *min max* *C1 7.0 < 20.0* *C2 20,0 < 50.0* *C3 50.0 < 95.0* *C4 95.0*
4.	20.8	13.4	7.4	
5.	20.9	12.5	8.4	
6.	22.5	14.4	8.1	

Assessment: the knitted fabrics tested fulfill the requirements for the resistance to the action of thermal radiation at an effectiveness level of C2.

Table 7. Results of testing the parameter of heat penetration with the action of flame

Sample No.	**Heat penetrating parameter HTI 24**	**Heat penetrating parameter HTI 12**	**HTI 24 – HTI 12**	***Requirements according to PN-EN ISO 11612:2011***
3.	11.0	8.0	3.0	*HTI 24* *min max* *B1 4.0 < 10.0* *B2 10.0 < 20.0* *B3 20.0*
4.	11.0	8.0	3.0	
5.	11.1	7.9	3.2	
6.	12.4	8.6	3.8	

Assessment: the knitted fabrics tested fulfill the required resistance to heat penetration with the action of flame at an effectiveness level of B2.

Table 8. Test results of the parameters of barrier properties concerning static electricity

Sample No.	Surface resistance R_s [Ω]		Volume resistance R_V[Ω]	Half decay time t_{50} [s]	Shielding coefficient S
	Fabric face	Fabric back			
3.	2.14 x 10^{12}	2.18 x 10^{12}	9.10 x 10^{10}	< 0.01	0.509
6.	4.72 x 10^{12}	1.66 x 10^{12}	9.02 x 10^{10}	< 0.01	0.59
Requirements according to PN-EN 1149-5:2009	$R_S \leq 2.5 \times 10^9$ *on at least one surface*		*1.66 x* 10^{12}	$t_{50} < 4s$	$S > 0.2$

Assessment: the knitted fabric tested show barrier properties concerning static electricity, fulfilling two parameters required, i.e. $t_{50} < 4$s. and $S > 0.2$

Table 9. The microbiological tests results of three-layer knitted fabrics (quantity method)

Sample No.	Micro-organisms used in tests according to: AATCC Test Method 100-2004 i PN-EN ISO 20743	Parameter determined	
		Value of the bacteriostatic parameter S	Value of the bactericidal parameter L
4.	Staphylococcus aureus (ATCC 6538)	2.2	1.1
	Escherichia coli (ATCC 11229)	-1.3	-2.1
5.	Staphylococcus aureus (ATCC 6538)	6.6	2.3
	Escherichia coli (ATCC 11229)	6.9	2.0
Requirements according to JIS L 1902		$S > 2$	$L \geq 0.$

Assessment: sample 4 shows bacteriostatic and bactericidal properties against bacteria Staphylococcus aureus, while against bacteria Escherichia coli this fabric has no biologically active properties. In the case of sample 5, its bacteriostatic and bactericidal properties were confirmed against both micro-organisms.

Table 10. Test results of the physiological comfort of three-layer knitted fabrics

Parameter \ Sample No.	3.	4.	5.	6.
Thermal resistance [m^2K/W]	0.083	0.070	0.072	0.090
Resistance of water vapor [m^2Pa/W]	2.87	6.01	6.11	7.35
Permeability of water vapor [g/m^2Pa h]	0.518	0.253	0.243	0.202
Permeability of air [mm/s]	1844	1766	1784	1328
Hygroscopicity [%]	8.91	9.89	9.97	9.90

Assessment: the values of thermal resistance indicate the thermal insulating power similar to that of fabrics used for warm outerwear (e.g. fabrics for coats – 0.093 m2 K/W) – assessment in relation to standard ISO 7730. The level of air permeability obtained, in relation to the classification of textiles, qualifies the knitted fabric as „air permeable" (according to standard PN-EN ISO 9237, the required level for the class "permeable" is over 600 m/s), while the level of water vapor permeability qualifies

IOP Conf. Series: Materials Science and Engineering **141** (2016) 012003 doi:10.1088/1757-899X/141/1/012003

the knitted fabrics tested to class 3 – "permeable" (according to EN 343:1996, class 3 – resistance ≤ 20 m^2Pa/W).

4. Summary

The technological and laboratory tests carried out at the Textile Research Institute indicate a chance of shaping the properties of multi-functional knitted fabrics with a three-layer structure and a specified use of them for clothing or its elements. The multi-functionality of these knitted fabrics shows protective properties against hot thermal effects and static electricity, and in the case of selected assortments, additionally biological activity against bacteria. At the same time, all the knitted fabrics tested are characterized by good parameters of physiological comfort.

The analysis and evaluation of indicators characterizing the multifunctionality of developed assortments of three-layer materials has allowed them to qualify as knitted fabrics with properties:

- flame-retardant and antistatic with good properties of physiological comfort – Sample 3 and 6,
- flame-retardant and antibacterial with good properties of physiological comfort – Sample 4 and 5.

The paper was prepared on the basis of the international project: Eureka E! 5799 BATAN „Barrier textiles and nanomaterials", co-financed by the National Center for Research and Development, Warsaw, Poland

References

[1] Information materials of Company Du Pont, USA.
[2] Information materials of SGL Group – The Carbon Company,
[3] Information materials of fibres and yarns producer Lenzing, Austria.
[4] Information material of Company Kuraray, Japan.
[5] Information material from Schoeller spinning group, Austria.
[6] Majchrzycka K, Poscik A 2007The selection of personal protective equipment, editor CIOP PIB, ISBN 978-83-7373-030-4
[7] Project No II-183/P-203/2007 „Active and biophysical clothing made from multilayer materials used in various climatic conditions", Textile Research Institute 2009
[8] Project No 6 ZR7 2008C/07115 "Technology of protective knitted fabrics against fire risk", Textile Research institute 2010
[9] Eureka project E! 5799 BATAN "Multifunctional knitted materials with barrier properties for clothing", Textile Research Institute, Lodz, 2014
[10] Mielicka E, Kozminska R, Janicka J, Walak A 2014 Multifunctional three-layer knitted fabrics, *conference material XI Seminar Research and Technical „Horizon 2020 The EU Framework Programme for Research and Innovation*, Ustron (Poland)
[11] Mielicka E, Janicka J, Kozminska R, Walak A, Oleksiewicz I 2014 Forming of properties of heat-resistant and anti-electrostatic double-layer knitwear, *Proc. of The 47th International Federation of Knitting Technologies Congress IFKT*, ISBN: 978-605-338-083-2, Izmir, Turkey

Comparison Of Flat-Knitted Structures Made Of Poly(P-Phenylene-2,6-Benzobisoxazole) And Para-Aramid Referring To Their Stab Resistance

M Obermann, S Aumann, F Heimlich, M O Weber, A Schwarz-Pfeiffer
Niederrhein University of Applied Sciences, Faculty of Textile and Clothing Technology, Mönchengladbach, Germany

E-mail: susanne.aumann@hs-niederrhein.de

Abstract. In the field of protective gear, developers always aim for lighter and more flexible material in order to increase the wearing comfort. Suppliers now work on knitted garments in the sports-sector as well as in workwear and protective gear for policemen or security-agents. In a recent project different knitted structures made of a poly(p-phenylene-2,6-benzobisoxazole) (PBO)-multifilament were compared to their counterparts made of para-aramid. In focus of the comparison stood the stab-resistance linked to either the mass per unit area or the stitch density. The tested fabrics were produced on hand flat knitting machines as well as on electronical flat knitting machines of the type Stoll CMS 330TC4, in order to analyse fabrics with different tightness factor and machine gauges. The stab resistance of the different knitted fabrics was examined according to the standard of the Association of Test Laboratories for Bullet, Stab or Pike Resistant Materials and Construction Standards. The presentation includes the depiction of the results of the test series and their interpretation. Furthermore it will give an outlook on most suitable combinations of materials and structures to be used in protective gear.

1. Introduction

Personal protective equipment is not only necessary for professionals, which are exposed to an increased risk of work-related potential, such as police officers, security guards, bodyguards, etc., but even in occupations where the risk potential does not appear obvious. In Germany, there are augmented reports of violence against people working in public jobs. Especially thrust weapons are used in such attacks. A practical protection of this vulnerable group of people appears more and more necessary [1,2]. Professionals like police officers are using special protective clothing (e.g. ballistic protection made of woven para-aramid or ultra-high-molecular-weight Polyethylene with integrated knife protection) [3]. In this application the protective effect is in focus while the wearing properties are getting secondary attention. Available on the market for civilians to protect themselves from knife attacks are mainly protective vests with integrated metal plates. Their highest disadvantage is the low wearing comfort due to their stiffness and high weight. In a recent project the Niederrhein University of Applied Science in cooperation with the knitting factory Bache Innovative developed a knitted jacket with integrated stab protection as a possible alternative [4,5]. In focus of this investigation which was inspired by a chain mail were not the protection requirements that apply to security officers, it was rather trying to find a compromise between sufficient stab protection and a comfortable fit. To reach this goal the advantages of knitted structures like the high flexibility and the breathability were combined with the properties of high-tensile polyethylene such as the low density and the high strength. To improve the stab resistance other materials have to be tested. A group of researchers also studied the stab resistance of multi-layered knitted fabrics made from para-aramid (poly-pentamethylen-perephthalat) and PBO (poly(p-phenylene-2,6-benzobisoxazole). They came to the conclusion that PBO-knitted fabrics are more suitable for textile stabbing-protection than PPT [6]. In the work described here, different knitted structures made of a PBO-multifilament were compared to their counterparts made of para-aramid.

2. Experimental

2.1. Knitting

Based on former experiments, four different structures were produced on an electronic flat-knitting machine of the type STOLL CMS 302 TC with gauge E8. Using Zylon AS (PBO) with a thickness of 555 dtex, the structures Milanese, French weave-knit, interlock and spacer-fabric were knitted. The same structures were manufactured using Twaron Type 2040 (para-aramid) with a thickness of 550 dtex (Figure 1). While knitting, the cut-resistant properties of the yarns led to difficulties, since the yarn would not break, when stretched too much, it would rather break the needles and bend the needle-bed. Due to this fact, the size of the loops got slightly increased. The speed of the fabric take-down was also decreased, since the used filaments are very sleek, so that the out-coming fabrics slipped through the cylinders. To minimize the risk of faulty stitches in the fabric, the knitting-machine was set on slow speed.

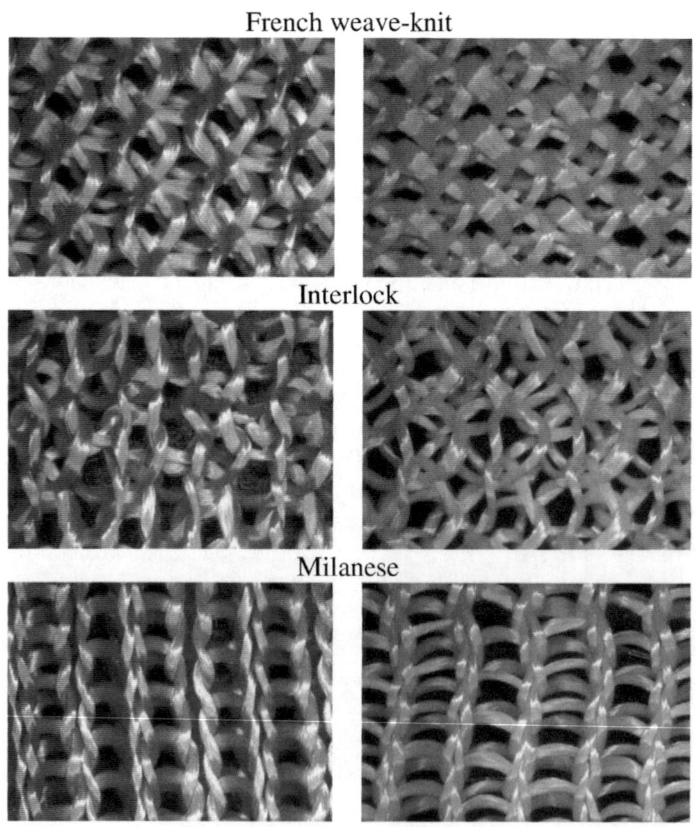

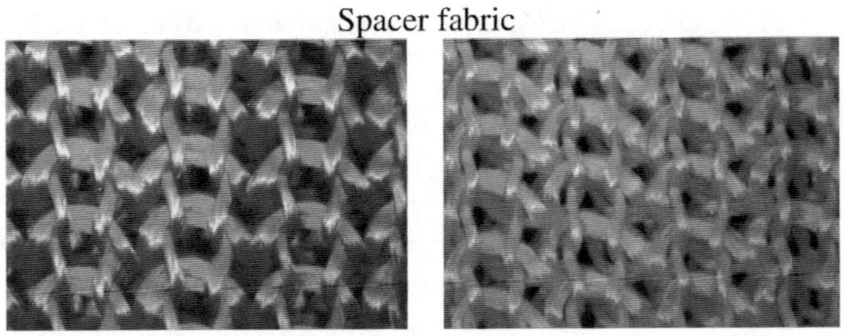

Figure 1. Microscopic pictures, 20 x zoom, PBO (left panels) and aramid (right panels).

While testing, it was observed that the PBO-fabrics developed peculiar craters that differed from the ones of aramid and UHMWPE (Ultra-high-molecular-weight polyethylene) fabrics (figure 2). They were steeper and deeper.

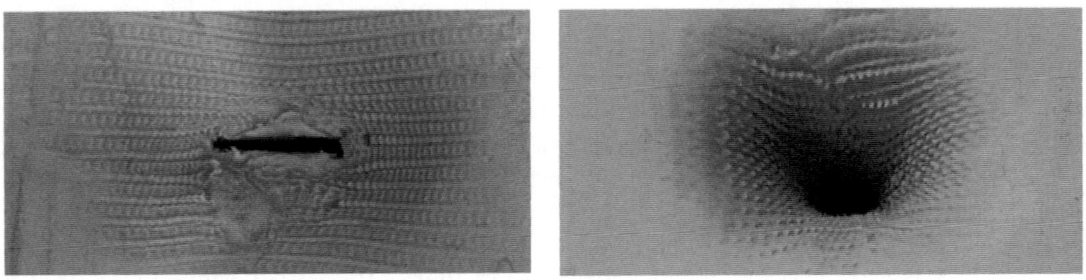

Figure 2. Cuts in plasticine, left panel: p-aramid as tested fabric, right panel: PBO as tested fabric.

To avoid this phenomenon in further fabrics, both filaments were used together in a spacer fabric. With the same machine properties two variations of the earlier-made spacer-construction were produced, where the spacer- and the loop-construction were made of either the aramid or the PBO-yarn, expecting them to compensate their counterparts' weak points, such as a deep crater.

2.2. Testing

The stab resistance of the different knitted fabrics was examined according to the standard of the Association of Test Laboratories for Bullet, Stab or Pike Resistant Materials and Construction Standards (VPAM 2011 [7]). According to this standard, protective gear is divided into different classes of protection. In protection class K 1 three tests with an impact-energy of 25 joule are run, in which every measured penetration depth must be less than 20 mm. This testing method checks textiles by letting a predefined blade with a predefined weight fall without braking out of a predefined height down onto the test-substrate, which is positioned on a plasticine-filled waggon. The weight onto which the blade is mounted is guided by two rail-bars, to ensure an impact in the right angle. It is fundamental, that the rail-bars guide the weight without decreasing its speed howsoever. The inner part of the waggon measures 350 mm x 400 mm x 150 mm. The ballistic putty has to be inserted in a way that prevents air-bubbles. Prior to every testing-session it has to be stored at a constant temperature that deviates by ± 2 °C at the most. The room temperature is stated at 20 °C (± 2 °C). The to-be-tested textiles should rest at least 12 hours at the same room temperature as well as at the relative air humidity of 65% (± 5%), which is the norm-climate. In order to test also thinner fabrics reliably, the falling height was reduced to 10 and 30 cm. To provide comparable results, the different fabrics were tested with varying numbers of layers to assure that their thickness would not differ much.

Table 1. Overview of yarns and the knitted structures produced from them.

Yarn parameters	Structures	Fabric thickness	Numbers of layers / overall thickness
Para-aramid Type 2040, 550 dtex	French weave-knit	2.62 mm	3 = 7.86 mm
	Interlock	2.73 mm	3 = 8.18 mm
	Milanese	2.66 mm	3 = 7.98 mm
	Spacer fabric	1.91 mm	4 = 7.64 mm
Zylon AS, 555 dtex	French weave-knit	2.03 mm	4 = 8.12 mm
	Interlock	3.58 mm	2 = 7.15 mm
	Milanese	1,97 mm	4 = 7.88 mm
	Spacer fabric	1.85 mm	4 = 7.40 mm
Zylon & Para-aramide	Zylon as tuck	2.37 mm	3 = 7.11 mm
	Aramid as tuck	1.72 mm	5 = 8.60 mm

3. Results

Figure 3 gives an overview of the results of the stab resistance tests of different knitted structures comparing PBO and aramid at a falling height of 10 cm (2.46 joule). The lower the result, the bigger is the stab resistance. Putting the PBO and the aramid knitted fabric to scale, the PBO always shows better results. The lower penetration depth of the PBO fabrics is significant in French knit-weave (PBO = 14.9 mm and aramid = 21.25 mm), Milanese (PBO = 12.79 mm and aramid = 22.18 mm) and spacer-fabric (PBO = 11.19 mm and aramid = 18.04 mm). The PBO and the aramid spacer fabrics have the best stab resistance results of all tested combinations (PBO = 2.47 mm and aramid = 2.7 mm).

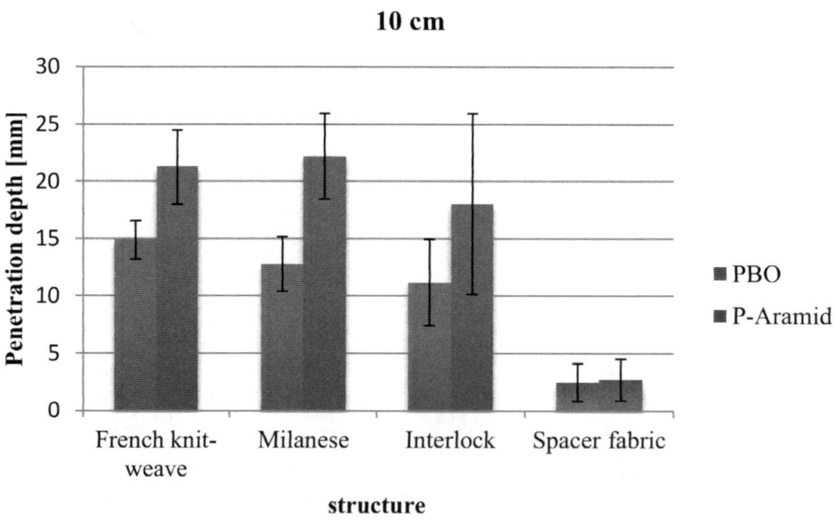

Figure 3. Comparison of the penetration depth of different knitted structures made of PBO and p-aramid (falling height: 10 cm).

Figure 4 shows the results of the same test and the same materials but with a falling height of 30 cm, which corresponds to an impact energy of 7.38 J. Again, the fabrics made of PBO achieve better results than the aramid fabrics. Only for the spacer fabric made from PBO the better result is significant (PBO = 12.03 mm and aramid = 19.52 mm).

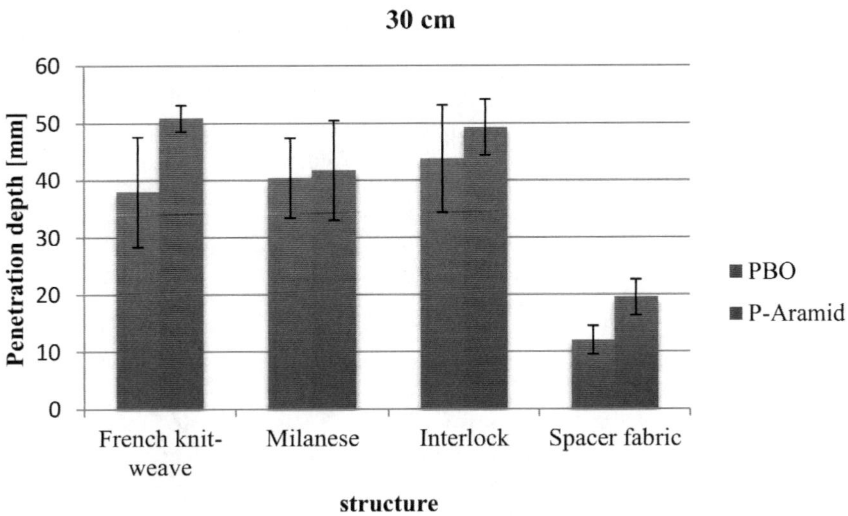

Figure 4. Comparison of the penetration depth of different knitted structures made of PBO and p-aramid (falling height: 30 cm).

Since the spacer fabric showed the best results in every comparison, figure 5 now puts the different material combinations to scale, but with the same knitted structure and with a falling height of 10 cm. The PBO fabric (2.47 mm), the aramid fabric (2.7 mm) and the PBO fabric with aramid used as tucks (2.83 mm) are showing almost the same stab resistance performance. The combined spacer-fabrics show a huge difference in their performance, since the data of the one with aramid used for the tucks is near its counterparts made of only one material while the construction that used PBO for the tucks has a remarkably higher penetration depth (10.66 mm). The craters of the combined spacers appeared lower than their PBO-counterpart as expected. Since this fact did not show any impact on the performance concerning the stab resistance, that was not yet investigated any further.

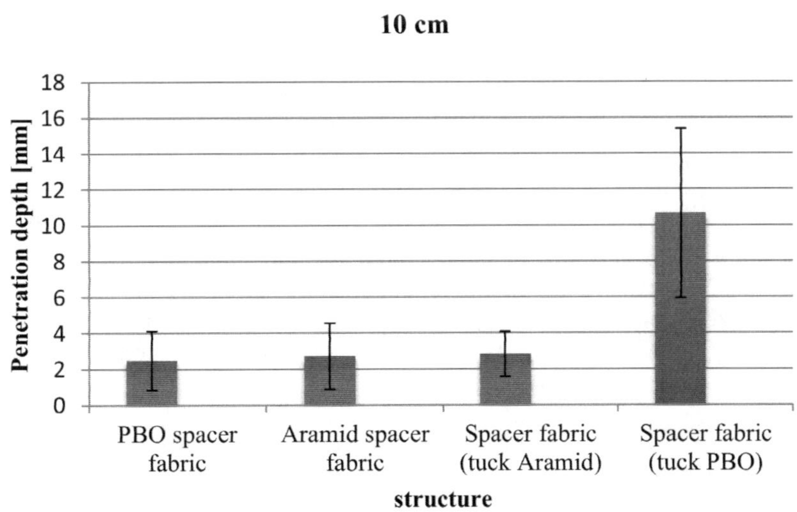

Figure 5. Comparison of the penetration depth of different spacer fabrics made of PBO and p-aramid (falling height: 10 cm).

Figure 6 shows the same fabrics as figure 5 only with an increased falling height of 30 cm. In this case the spacer-fabric made of both aramid and PBO with aramid used in the tucks shows the best resisting-performance (10.68 mm). Regarding the standard deviation, the stab resistance of the PBO with aramid in the tucks is in the same area as the PBO spacer fabric (12.04 mm). Both have a better performance than the aramid spacer fabric (19.52 mm). The combined spacer-fabric that uses PBO for the tucks has similarly insufficient results (26.16 mm) as in figure 5.

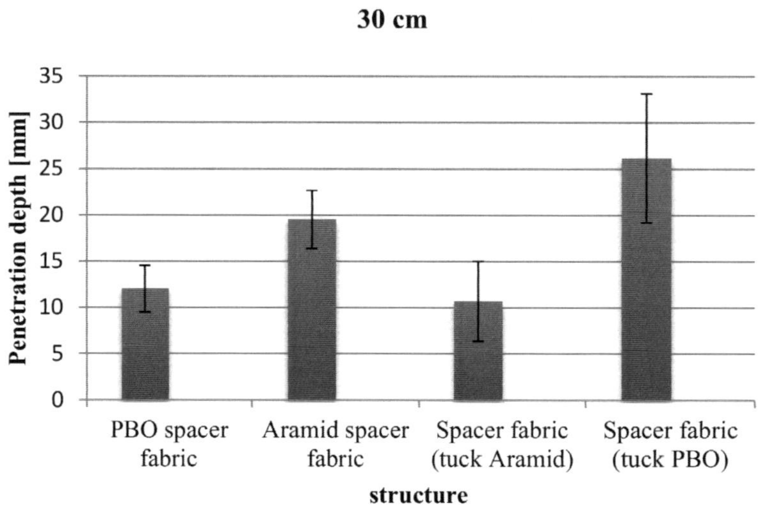

Figure 6. Comparison of the penetration depth of different spacer fabrics made of PBO and p-aramid (falling height: 30 cm).

Due to the inhomogeneous structures of knitted fabrics, the test results have relatively high standard deviations. The spacer fabric made from PBO has the lowest standard deviation which can be attributed to the fact that its structure is more even and stable (Figure 1).

4. Conclusion and Outlook

After having revisited numerous combinations of material- as well as construction-parameters the spacer-fabrics made of single PBO and PBO with aramid used as tucks seem to be the most promising. There will be further tests on material combinations, starting with combining PBO and UHMWPE in a spacer fabric, as well as spinning the three materials together and using the out-coming yarn for the structures mentioned above. Additionally new constructions will be tested with the PBO-fiber to decrease the flexibility of the single loops, so that the blade can be caught more easily. Last but not least there will be run further tests on the PBO-spacer fabric used above, such as martindale and washing-tests to ensure its suitability for the use in protective garments. To assure a better comparability and to minimize the craters the upcoming fabrics will be tested on top of a non-woven, which is the testing manner of the HOSDB, which provides the standard for body armor for the UK police.

5. References

[1] LN Online. Messerangriff auf Busfahrer: Die Angst fährt immer mit [Online] http://www.ln-online.de/Lokales/Luebeck/Kuecknitz/Messerangriffauf-Busfahrer-Die-Angst-faehrt-immer-mit [08.09.2015]

[2] WDR. Tödliche Messerattacke in Jobcenter, [online] http://www1.wdr.de/themen/panorama/jobcenterneuss106.html [22.05.2015]

[3] BGZ – Hauptverband der gewerblichen Berufsgenossenschaften Fachausschuss Persönliche Schutzausrüstungen. BGR 196 – Benutzung von Stechschutzbekleidung, [online] http://publikationen.dguv.de/dguv/pdf/10002/bgr196.pdf [20.05.2015]

[4] Aumann S, Weber M, Ehrmann A 2015 Entwicklung einer Strickjacke mit Stich- und Schnittschutzwirkung, Konferenz MG Open Spaces – Forschung aktuell

[5] Vidzem B, Aumann S, Heimlich F, Werner R, Ehrmann A, Obermann M, Brücken A, Weber M O, Bache T 2013 Stab resistance of textile materials, *Aachen-Dresden International Textile Conference: book of proceedings*, Aachen: November 2013

[6] Flambard X, Polo J 2004 Stab resistance of multi-layers knitted structures: comparison between para-aramid and PBO fibers, *Journal of Advanced Materials* **36** 30-35.

[7] N. N. 2011 Vereinigung der Prüfstellen für angriffshemmende Materialien und Konstruktionen (VPAM). Prüfrichtlinie Stich- und Schlagschutz, VPAM-KDIW 2004

Acknowledgment

This project is funded by the Bundesministerium für Wirtschaft und Energie due to a decision of the Deutscher Bundestag.

Systematic development of technical textiles

M Beer[1], V Schrank[1], Y-S Gloy[1] and T Gries[1]
[1] Institut für Textiltechnik (ITA) der RWTH Aachen University, Aachen, Germany

E-mail: mathias.beer@ita.rwth-aachen.de

Abstract. Technical textiles are used in various fields of applications, ranging from small scale (e.g. medical applications) to large scale products (e.g. aerospace applications). The development of new products is often complex and time consuming, due to multiple interacting parameters. These interacting parameters are production process related and also a result of the textile structure and used material. A huge number of iteration steps are necessary to adjust the process parameter to finalize the new fabric structure. A design method is developed to support the systematic development of technical textiles and to reduce iteration steps. The design method is subdivided into six steps, starting from the identification of the requirements. The fabric characteristics vary depending on the field of application. If possible, benchmarks are tested. A suitable fabric production technology needs to be selected. The aim of the method is to support a development team within the technology selection without restricting the textile developer. After a suitable technology is selected, the transformation and correlation between input and output parameters follows. This generates the information for the production of the structure. Afterwards, the first prototype can be produced and tested. The resulting characteristics are compared with the initial product requirements.

1. Introduction

Textiles can be used in various fields of applications, e.g. the automotive sector, sports industries medicine and apparel. The implementation of new textile product ideas is hindered due to the variety of usable machinery and process parameters. A suitable design method for the development and production of new technical fabrics can be used as a guideline for the systematic realisation of these ideas. The method helps to link different views; know-how of technicians, customer wishes and machine possibilities are all take into consideration to avoid misunderstandings or limitations. A new, successful development can only be achieved in close collaboration between these different disciplines (Figure 1.) [1]

General design methods for the systematic development and production of new products already exist [2, 3, 4, 5]. These existing methods are laid out for product development without link to a specific technology field which makes the transfer into the textile industry difficult. Therefore, a new approach is presented that includes the different textile fabric technologies.

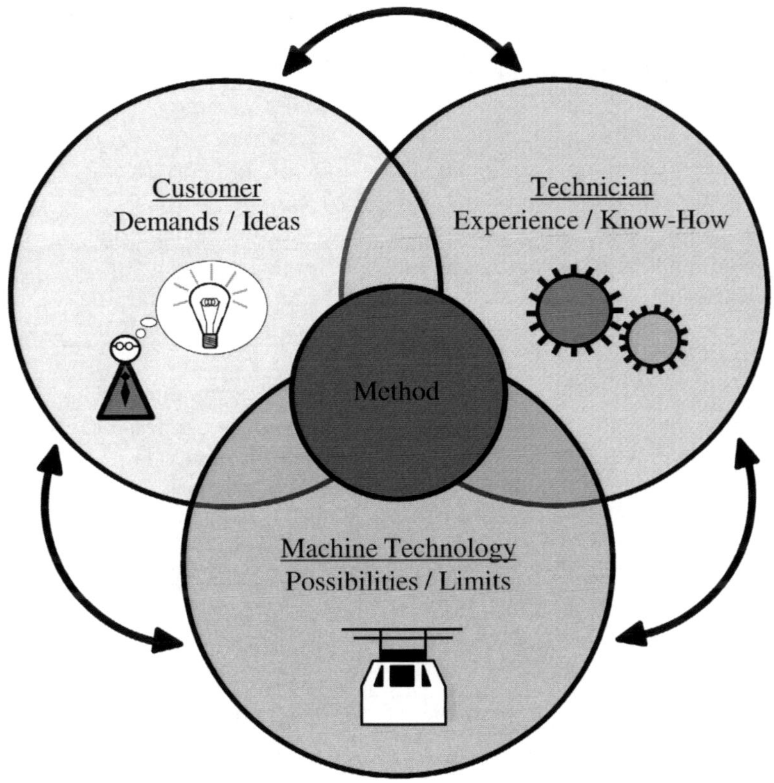

Figure 1. Linking of different perspectives with the help of a development method for technical textiles.

2. Methods of product development

A new approach is presented within this paper for the systematic development of technical textiles. Following, an overview of different, currently existing methodologies and tools for the product development in different industries and development steps are described. These methodologies contain methods to support the development process, methods for the general proceeding during the product development and construction methodologies. The deficit of these methodologies for the development of textile structures is shown.

2.1 Classification of current methods

Supporting methods for the product development are used in various steps during the development process. The following overview shows exemplarily methods for different application areas. A possible classification of supporting methods is given. Often, these supporting methods can be applied for multiple product development steps.

Table 1. Overview of different methods and classification thereof.

Classification / filed of application	Method
Methods to generate ideas	• Brainstorming • Method 635 • Delphi expert survey • TRIZ • Construction catalogues
Methods to identify requirements & documentation	• Survey/questionnaire • Product-benchmarking • Specification sheet
Evaluation methods	• Kano-model • Pairwise comparison • Strength/weakness profile • Portfolio-analysis • Utility analysis • Check lists • Morphological box
Methods for evaluation of economic efficiency	• Cost-benefit analysis • Machine hourly rate calculation • Amortisation calculation • Break-Even-Analyse
Methods for design of experiments	• Design of Experiments (DoE) • Factorial DoEs

2.2 General product development methods

The steps during the development process of new products are similar independent on the market and field of application. The main part of each development process is the planning of the product, the product conception and the product design (Figure 2.). [2]

During the product planning the requirements are defined. Customer requirements and competitor products are systematically analyzed and an economic analysis is made. The requirements should be quantized and generally 10 – 20 requirements are sufficient to fulfill all requirements. The product conception contains the search for possible solutions for the new product. With the help of assessment criteria the most suitable solutions are chosen. Within this product development phase the suitable technology for the production needs to be identified. The final phase, the product design, contains the actual construction of the product. [2] During all these steps of the product development process, the supporting methods of Table 1 are used.

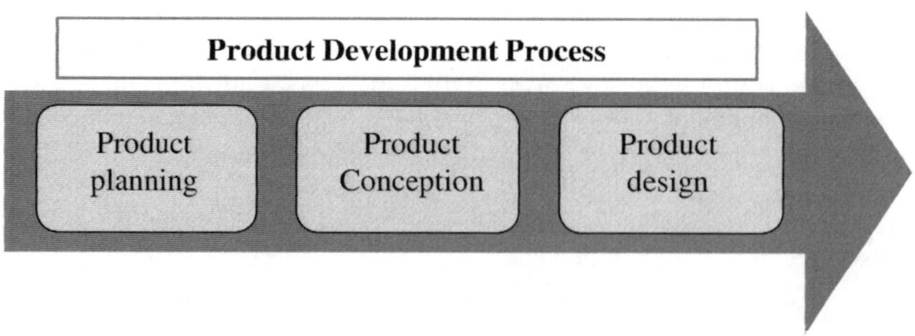

Figure 2. Steps of common product development methods.

2.3 Deficit of current methods

Several product development methods already exist (Figure 3). Some of these methods can be universally used. These methods can be transferred to all kinds of product development processes (e.g. Quality Function Deployment (QFD), VDI Guideline 2221). Other methods have been especially developed for textile applications or to support the textile production (e.g. 9-Step-Tool, Textile Parametric Design (TPD)) [2, 6, 7, 8, 9]

A universally usable methodology for the development of technical textiles does not exist. The textile methodologies all focus on specific product categories or technologies. To ease the development of new textile products and to accelerate the market entry for companies in a new sector a universally usable textile methodology is needed. This methodology focuses especially on textiles. A main part of the method is the identification of a suitable production technology.

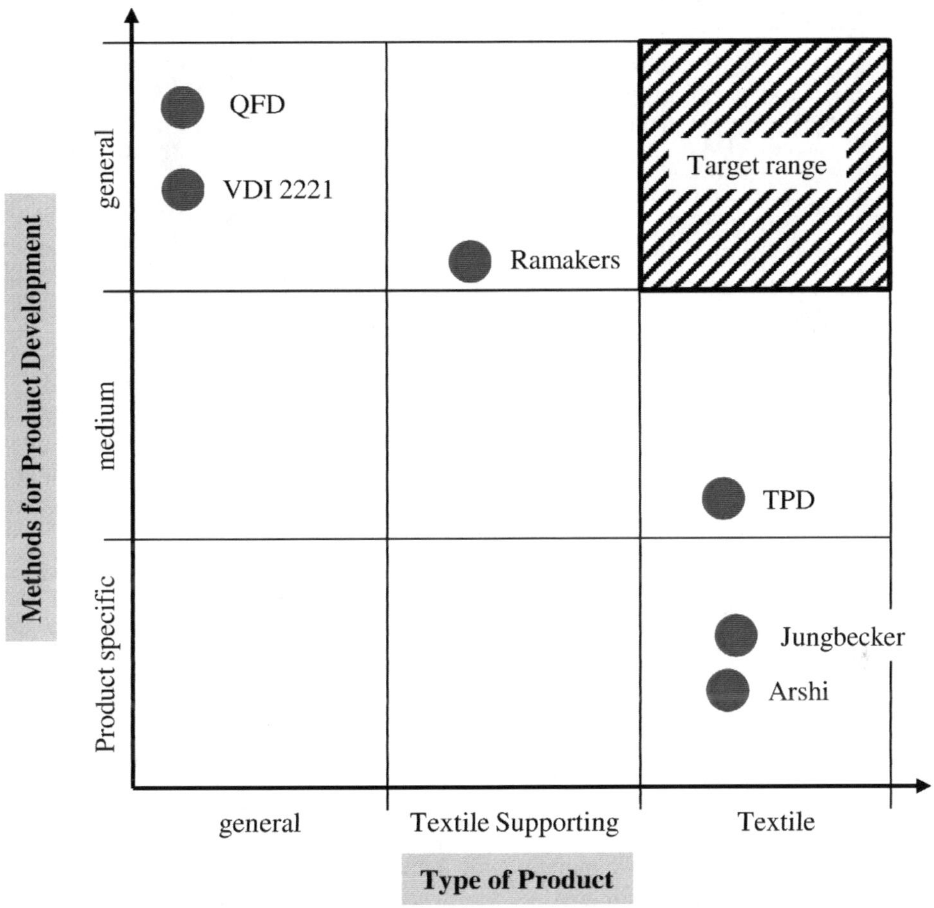

Figure 3. Portfolio illustration of different product development methods.

3. TED Method – Universal Textile product development method

The developed and proposed TED Method (**TE**xtile **D**evelopment Method) is subdivided into six steps (Figure 4). Following, these six steps from requirement definition to the validation of the new product are described. The steps are described without link to a specific textile product and its requirements to show the general proceeding within the method.

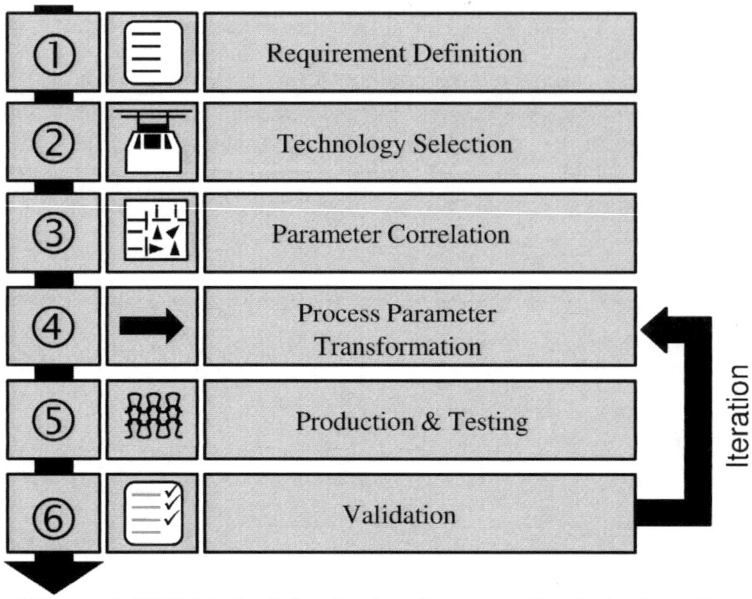

Figure 4. TED Method for the development of technical textiles.

3.1 Step 1 - Requirement Definition

The identification of the requirements of a new textile product is essential for the development thereof. The fabric characteristics vary depending on the field of application. Therefore, the first step is the identification of the desired requirements. The definition of requirements is further subdivided into the following steps:

- collection of all requirements / benchmark test
- ranking of requirements
- transfer into measurable fabric parameters
- value allocation – defined, prioritized requirements list

Overall requirements are collected within a workshop (brainstorming, clustering and ranking) with experts of the development team. Terms for the overall requirements are e.g. soft, strong and good climate comfort. These general requirements are often non-measurable properties. Therefore, if a benchmark or similar product exists, this product is tested to identify measurable properties. The collected requirements are ranked to identify the most important characteristics. Some properties can be contradictory like air permeable and water-repellent. After the ranking or the requirements, they are transferred into measurable fabric characteristics. Non-measurable characteristics can be correlated to multiple measurable characteristics and test standards (e.g. fabric breathability = air permeability, fabric density). For the validation after production, target values of the characteristics are needed.

3.2 Step 2 - Technology Selection

A suitable technology needs to be selected depending on the requirements of the desired product in development. Fabric production technologies vary from weaving, weft and warp knitting, braiding and non-woven production to multi non-crimp fabrics. All these technologies can be further sub divided dependent on resulting dimensions (2D, 3D), structure (velvet, tubular, etc.), configuration (weft insertion technology, weft and warp yarns, plush, etc.). Moreover, the productivity and economic efficiency are taken into consideration for the technology decision. [10, 11]

Aim of this method step is to outline all possible production technologies applicable for the desired product. This allows broadening the scope within the development process. Finally, the best fitting technology solution for the application will be selected, to follow within the next method steps. To build a robust technology selection tool the method of a construction catalogue [12] is used. This construction catalogue is a data collection of all market available fabric production technology solutions.

3.3 Step 3 - Parameter Correlation

The requirements, gathered in method step 1, define textile characteristics of the final fabric. These characteristics are the targeted output parameters of the production process. Up to this point, the input parameters of the production process are not known. The input parameters (such as material, yarn tension, machine velocity and pattern) are the determining factors for the output characteristics. Input and output parameters need to be correlated based on a Design of Experiment (DoE) study to choose the appropriate input (material, machine setting, pattern) for the requested output (fabric requirements). This correlation of parameters is carried out in method step 3. Figure 5 shows the necessary transfer link from product requirements into machine input parameters.

First, the production technology specific process input parameters are identified. Depending on the selected technology, different process parameters can be changed. All input parameters can be divided into three groups: material, structure/pattern, machine setting.

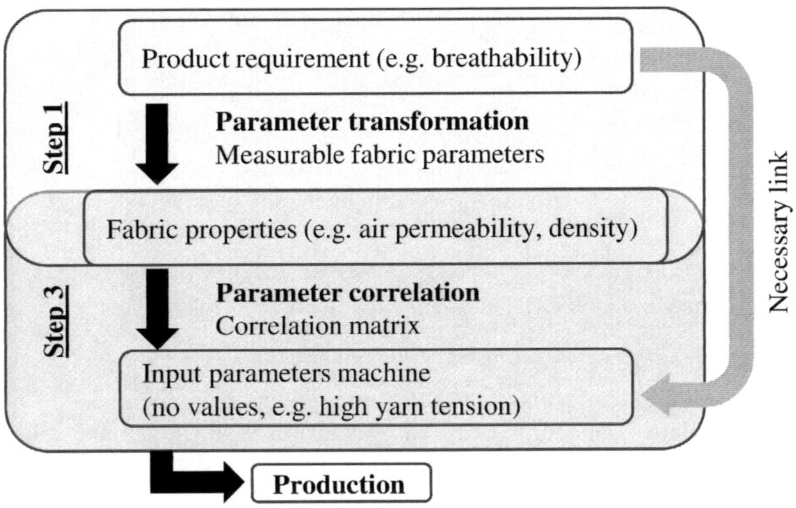

Figure 5. Transfer from product requirements into machine input parameters.

A correlation matrix is needed to correlate the input and output parameters. The correlation effects have to be determined within pre-trials. One example is a knitted product where the air permeability is required to be as high as possible (Table 2). The given correlation matrix shows, that a low machine gauge is suitable to archive this fabric characteristic. Moreover, the number of used yarns per system should be as low as possible as well as the yarn tension. The sinking of the loops influences the stitch size of the mesh. The stitch size should be high, to produce fabrics with an open structure. The correlation effects are moreover shown in corresponding diagrams.

Table 2. Exemplary correlation matrix.

	Input parameter - Machine setting			
Fabric property	Machine gauge ↑	No. of feed yarns per needle ↑	Yarn tension ↑	Sinking (stitch size) ↑
Air permeability	↓	↓	↓	↑

3.4 Step 4 - Process parameter transformation

The defined parameters need to be transferred into machine data. The production parameters that influence the resulting fabric properties have been identified within method step 3. Within method step 4, the relevant parameters are transferred into measurable values with the help of the tendencies from Table 2. Input parameters such as material, structure/pattern and machine setting are dependent on the used textile production machinery.

Value ranges for all parameters are given by the technology specific limits. For example for the production of highly breathable knitted fabrics, the yarn tension is varied. Exemplary for a selected machine, the tension can be adjusted from 1 cN to 20 cN. The tendencies from the correlation matrix are multiplied by the influencing amount (air breathability is influenced by multiple parameters (e.g. machine gauge, no. of yarns, stitch size). The resulting value is calculated (Figure 6).

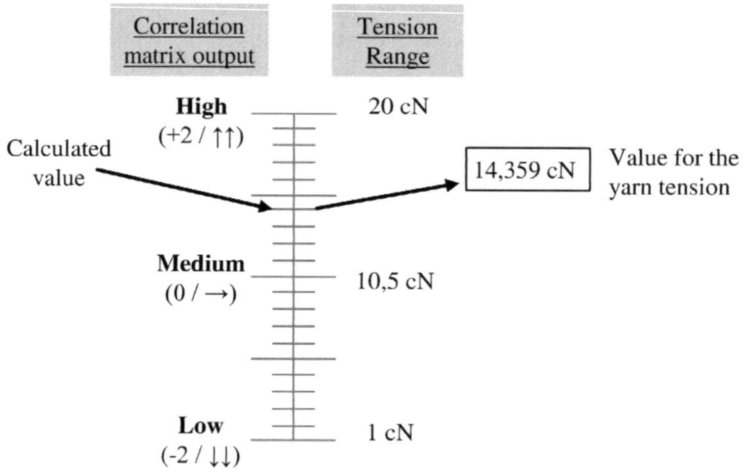

Figure 6. Transfer from correlation tendencies into values.

3.5 Step 5 - Production & Testing

Production as well as testing is carried out in method step 5. The machine needs to be setup for the production of the new product. The setup is carried out according to the defined process parameters from method step 4. After the set-up phase, the machine needs to produce as much material as necessary for the laboratory tests. For standard fabric tests (stress/strain, air permeability, thickness and mass per unit area) about three square meters of fabric are necessary.

The produced material samples are tested in a laboratory regarding their mechanical properties. The testing standards have already been defined in method step 1 and are linked to each characteristic in the requirement document. The results of this sub-step are numerical fabric characteristics, such as air permeability and mesh density.

3.6 Step 6 - Validation

The final step within the design method is the validation. The produced material needs to be evaluated regarding the fulfillment of the requirements from the specification sheet. The first step is the review of the material sample results. The tested fabric properties (laboratory) are compared to the defined requirements (requirements list, Figure 7). If necessary, an adjustment loop is carried out with adapted process parameters based on the correlation tendencies. For this, the design method is used once more, starting from method step 4.

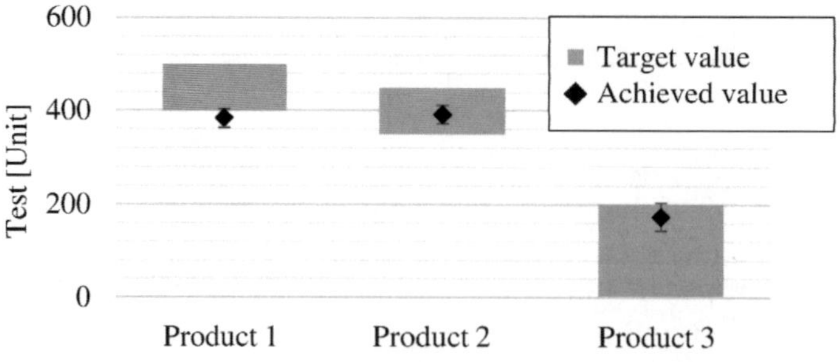

Figure 7. Comparison: target and achieved value.

4. Summary

To perform a systematic development of new products, several general method approaches are available on the market. Before though, there was no specific method for the development of technical textiles. To simplify the approach of developing new textile products a new method was developed and presented – The TED Method (TExtile Development Method). The TED Method was validated at the Institut für Textiltechnik (ITA) der RWTH Aachen University, Aachen, Germany for two different knitted products from different application categories, a functional apparel component and a technical automotive textile. It was shown that for both applications, the development iteration steps using the design method could be reduced drastically to a maximum of two overall iterations per product. The core element of the design method to reduce the number of needed iterations is the parameter correlation, based on detailed material trials. The next step at ITA to further develop the method will be to establish parameter correlations for each common fabric production technology.

5. References

[1] Beer M, Simonis K, Schrank V, Gloy Y-S, Gries T 2014 *Innovation in knitting technologies and applications* (Izmir, Turkey, IFKT 2014 Conference)

[2] Engeln W 2004 *Methoden der Produktenwicklung* (Pforzheim: Pforzheim University of Applied Science)

[3] Kröll M 2007 *Methoden zur Technologiebewertung für eine ergebnisorientierte Produktentwicklung* (Stuttgart: Universität Stuttgart)

[4] Schuh G, Klappert S 2011 *Technologiemanagement – Handbuch Produktion und Management 2* vol. 2 (Berlin; Heidelberg: Springer Verlag)

[5] VDI-Richtlinie 3780 2000 *Technikbewertung – Begriffe und Grundlagen* (Düsseldorf: Beuth-Verlag Berlin)

[6] VDI Richtlinie 2221 1993 *Methodik zum Entwickeln und Konstruieren technischer Systeme und Produkte* (Düsseldorf : Beuth-Verlag Berlin)

[7] Ramakers R 2005 *Systematische Entwicklung von sensorbasierten Online-Überwachungssystemen für die Filamentgarnverarbeitung* (Aachen, Techn. Hochsch., Diss.; Zugl Aachen : Shaker)

[8] Sri Harwoko M 2008 *Entwicklung einer Methodik zur systematischen Gestaltung von Gewirken für den Einsatz als textile Implantate* (Aachen, Techn. Hochsch., Diss.; Zugl. Aachen : Shaker)

[9] Jungbecker N *2011 Gestaltung und Charakterisierung textiler Elektroden zur Überwachung von Körperfunktionen* (Aachen, Techn. Hochsch., Diss.; Zugl. Aachen : Shaker)

[10] Beer M, Schrank V, Gloy Y-S, Gries T 2015 *Production technology selection for the development of technical fabrics* (Geelong, Deakin University, ATC-13 Asian Textile Conference)

[11] Schrank V, Beer M, Gloy Y-S, Gries T 2015 *Technology selection for innovative products* (Aachen, DWI Leibnitz-Institut, 9th Aachen-Dresden International Textile Conference)

[12] VDI Guideline 2222 Sheet 2 1982 *Konstruktionsmethodik – Erstellung und Anwendung von Konstruktionskatalogen* (Düsseldorf: Beuth Verlag Berlin)

IOP Conf. Series: Materials Science and Engineering **141** (2016) 012006 doi:10.1088/1757-899X/141/1/012006

Influence of miss knit repeat on parameters and properties of elasticized knitted fabric

O Kyzymchuk, L Melnyk

Kyiv National University of Technologies and Design,
2 Nemirovych-Danchenko str., Kyiv, UKRAINE

kyzymchuk.o@knutd.com.ua; kyzymchuk_knutd@ukr.net

Abstract. The features of elasticized knitted fabric are: high extensibility, elasticity and adaptability to a body shape and a compression effect when it is used in the extended state. The elasticity of a knitted fabric is enhanced by the incorporation of elastane thread or core-spun yarn into the knitted structure as the weft yarn that is laid in the stretching direction. The quality of elasticized knitted fabrics is determined by the reliability of the how the elastomeric yarn is fixed within as the structure is deformed with an increase in the contact areas between the elastomeric and conventional yarns. The high fixing level can be achieved by the nature of loops formed from the elastomeric yarn. The simultaneous union of both conditions is possible with miss-knitting of elastomeric yarn through a certain number of needles. The main purpose of this study is to establish the influence of miss knit repeat of elastomeric yarn on the structural parameters and determine the stretch properties of the elasticized knitted fabric that formed by an alternation of two courses of 1x1 rib from cotton yarn and one course of miss knit (from 2 to 10 wales between the loops) from core-spun yarn with elastane core.

1. Introduction

Application of textiles in medicine and health care can prevent and delay disease progression. The knitted fabrics with incorporated elastane is one of textile materials which use has increased in recent decades. The features of such materials include high extensibility, elasticity and adaptability to a body shape that is indispensable in compression garment [1]. In addition, elasticized knitted fabric creates a compression effect when it is used in the extended state [2]. That is why the elasticized knitted fabrics are indispensable in the manufacturing of therapeutic and prophylactic products, such as elastic bandages. bands for different purposes and abdominal binders [3]. The quality, functional and hygienic properties of materials are very important for the manufacture of medical items for rehab assignment. The hygienic requirements are ensured by using knitted fabrics from natural yarn. The physical and mechanical properties are achieved through the use of elastomeric yarn.

The elasticity of a knitted fabric is provided by the incorporation of elastane thread or core-spun yarn with elastane core into the knitted structure as the filling yarn that is laying in the stretching direction [4]. For example, when the elastomeric yarns are used as weft-inserted thread into the weft knit, the fabrics has got an extensibility in coursewise direction. At the same time, an elastomeric yarn should not be placed at the surface of knitted structure to provide the necessary hygienic properties. The elastomeric yarns position inside the knitted structure can be provided by using the double fabric such rib 1x1 as a base, wherein the filling yarn is disposed between the face and reverse loops (Fig.1.). The special yarn feeder is required on the knitting machine used for the manufacture of such weft knitted fabric. This yarn feeder is located in front of the system. which is forming basic interlooping. The filling yarn is laid between the needle beds when the needle heads of both beds are disposed below the knock-over plane.

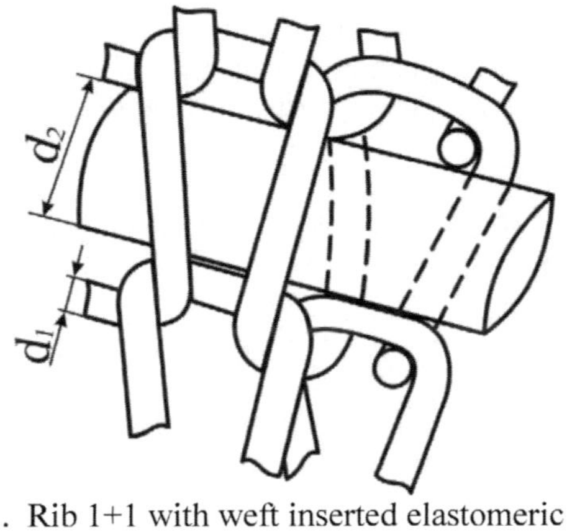

Figure 1. Rib 1+1 with weft inserted elastomeric yarn

At the same time, the quality of materials with incorporated elastomeric yarn is determined by the reliability of the elastane fixing in the structure that is increasing with an increase – in the contact areas between the elastomeric and conventional yarns. This can be achieved by increasing the density of knitting, when the loop height does not exceed the diameter of the filling yarn. However, there would be a problem with relaxation of the elastomeric thread in this structure, that would significantly affect the extensibility and elasticity of the fabric.

From the other point, the high fixing level of the elastomeric yarn can be achieved by the loops formation from this yarn [5]. This method is now widely used in the production of sports fabrics and linen from core-spun yarn with elastomeric core [6]. When using polyurethane without any wrapping the hygiene requirements to fabric are violated.

The simultaneous union of conditions of elasticity, hygiene and reliability of fixing of the elastomeric yarn is possible with miss knitting of elastomeric yarn through a certain number of needles in a certain courses. The elastomeric yarn is deployed as weft filling yarn in a certain number of missed wales and is connecting the formed loops.

2. Theoretical Consideration

It is possible to knit elastomeric yarn separately [7] or together [8] with conventional yarn. When loops are formed from both together elastomeric and conventional yarn. the elastomeric yarn is positioned inside the knitted structure. The loops from elastomeric yarn create an elastic frame between elements from conventional yarn. The loops' skeletons from conventional yarn are changing their position to coursewise direction. They are bending and are turning in a plane that is perpendicular to the fabric plane [9]. But when loops are formed from elastomeric and conventional yarn together, accordingly they are laid and draw-off simultaneously. In this case, the properties of knitted fabric are determined by conventional yarn: the elongation of the fabric is limited by junctures from conventional yarn that are most orientated in the stretching direction.

The elasticized fabric produced at separate knitting of elastomeric and conventional yarns are different. The elastomeric yarn is trying to regain its former size and to reduce its length after knitting and the termination of draw-off force. As a result, knitted fabric has got high elasticity, in this case the elastic properties depend on the pretension of the elastomeric yarn before knitting and on the degree of relaxation in the knitted structure [10].

This fabric is produced on double V-bed flat knitting machine equipped with a needle selection mechanism. The conventional yarn 1 is laid in the first knitting system and forms the loops of rib 1x1 (Fig. 2). The elastomeric yarn 2 is laid in the second knitting system with pre-elongation and forms the loops depending on the repeat. The elastomeric yarn can form loops by needles as only one needle bed (Fig. 2.a), or both needle beds (Fig. 2.b-d). One (Fig. 2.a-b) or two (Fig. 2.c-d) elastomeric yarn can be laid at one course, the elastomeric yarn can be laid in each course or according to a particular repeat. It`s possible to form the loops from elastomeric yarn on adjacent needles (Fig. 2.d), and with offset (Fig. 2.c) on different needle beds.

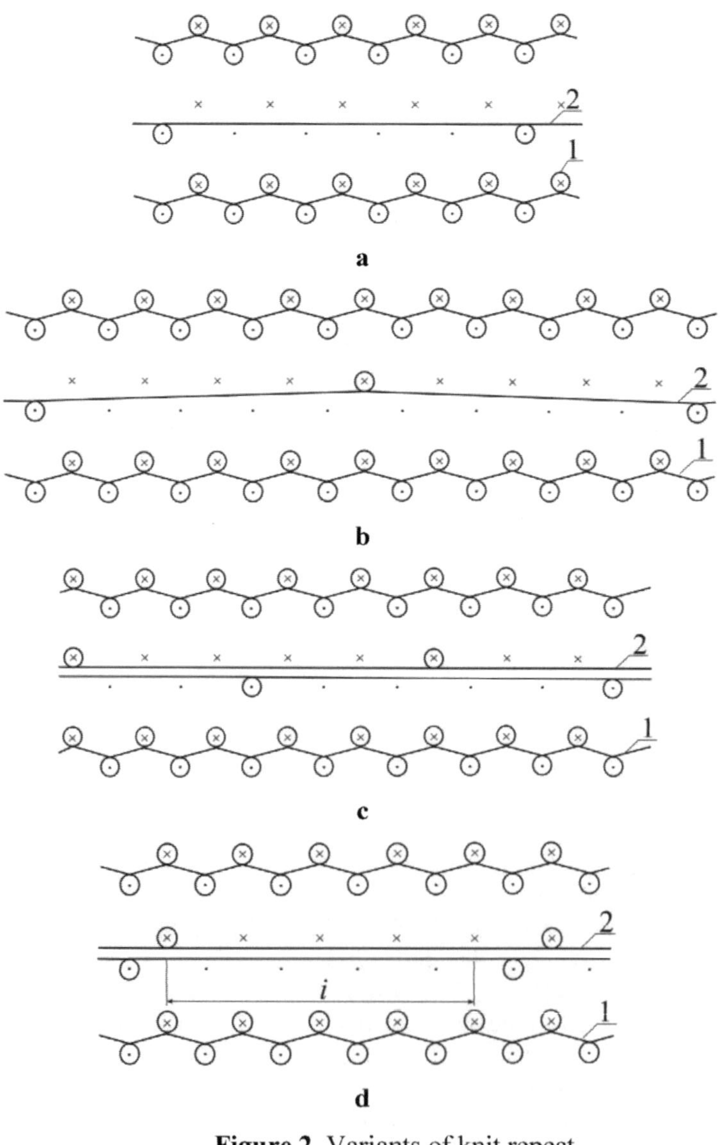

Figure 2. Variants of knit repeat

It is obvious that a preliminary elongation of the elastomeric yarn and the fabric density are the main factors that determine the structure parameters and mechanical properties of elasticized knitted fabrics [11]. However, it should be assumed that laid repeat [12] and miss knit repeat of elastomeric yarn will also affect the performance properties of knitted fabrics.

Thus, the main purpose of this study is to establish the influence of miss knit repeat of elastomeric yarn on the structure parameters and the stretch properties of elasticized knitted fabric.

3. Experimental Parameters

The weft knitted fabric that was formed by an alternation 2 courses of 1x1 rib from cotton yarn and 1 course of miss knit from core-spun yarn with elastane core has been chosen for this study. The loops from elastomeric yarn are formed by adjacent needles on different needle beds (Fig. 2.d). In order to verify the effect of miss knit repeat of the elastomeric yarn on the performance properties of the fabric, the number of missed wales between the loops from the elastomeric yarn has been varied from 2 to 10 with an interval of 2.

Experimental samples of knitted fabric have been produced on 10-gauge flat V-bed knitting machine. The weft knit fabrics were made from 31 tex * 2 cotton yarn as ground. The core-spun yarn with 90 tex polyurethane yarn covered by two end of 12.5 tex polyester yarn was used as elastomeric yarn. Elastomeric yarn was fed into the knitting zone with elongation of 145%. that made it possible to obtain high-quality fabric during loop formation.

The investigation of structure parameters and of strain characteristics of elasticized fabric was carried out according to standard methods after 72-hour relaxation in normal climatic conditions. The experimental results are presented in Table 1.

Table 1. Experimental Data

			Number of miss needles at the repeat, *i*				
			2	4	6	8	10
Length, mm	of loop	l_l	5.55	5.55	5.57	5.60	5.59
	of weft	l_2	0.84	0.83	0.81	0.80	0.79
Number per 100 mm	of courses	N_c	76	75	74	73	72
	of wales	N_w	86	88	92	96	98
Thickness, mm		T	2.20	2.16	2.10	2.06	2.01
Basic weight, g/m^2		m_s	658.18	664.34	669.98	674.48	677.78
Full stretch, %		ε	49.0	81.4	104.6	126.4	148.4
Parts of full stretch, %	elastic	ε_1	45.4	77.0	97.0	116.8	137.6
	plastic	ε_2	2.6	3.0	4.0	5.2	6.2
	residual	ε_3	1.0	1.4	3.6	4.4	4.6
Quota of part	elastic	$\Delta\varepsilon_1$	0.927	0.946	0.927	0.924	0.927
	plastic	$\Delta\varepsilon_2$	0.053	0.037	0.038	0.041	0.042
	residual	$\Delta\varepsilon_3$	0.020	0.017	0.034	0.035	0.031

4. Analysis

The single-factor mathematical models that describe the influence of missed knit repeat *(i)* of elastomeric yarn on the studied parameters were obtained as a result of mathematical calculation and are presented in Table 2. All dependencies are linear.

Table 2. The Mathematical models

№	Parameter	Equals
1	Length of elastomeric weft yarn, mm	$l_2 = -0.006\,i + 0.845 \quad (R^2=0.81)$
2	Number of courses per 100 mm	$N_c = -0.4\,i + 76.5 \quad (R^2=1)$
3	Number of wales per 100 mm	$N_w = 1.6\,i + 82.4 \quad (R^2=0.98)$
4	Thickness, mm	$T = -0.02\,i + 2.25 \quad (R^2=0.99)$
5	Basic weight, g/m^2	$m_s = 2.47\,i + 654.15 \quad (R^2=0.98)$
6	Full stretch, %	$\varepsilon = 12.19\,i + 28.82 \quad (R^2=0.99)$
7	Elastic stretch, %	$\varepsilon_1 = 11.21\,i + 27.50 \quad (R^2=0.99)$
8	Plastic stretch, %	$\varepsilon_2 = 0.47\,i + 1.38 \quad (R^2=0.97)$
9	Residual stretch, %	$\varepsilon_3 = 0.51\,i + 0.06 \quad (R^2=0.90)$

4.1. Structural parameters

It is well known that yarn tension and force of fabric draw-off as well as draw-off level (knock-over depth) are the main factors that affect the loop length of weft knitted fabric. The loop length of rib 1x1 (l_1) is not affected by miss knit repeat of elastomeric yarn as all experimental samples are made on the same knitting machine at constant knitting parameters. At the same time, it was found that the average length of the elastomeric weft yarn (l_2) is reduced by increasing the number of missed wales. Such direct proportion is due to the difference in lengths of the loop and portion of weft yarn that connected two loops. So the length of weft yarn in fabric with 2 missed wales is two loop pitches and the length of weft yarn in fabric with 10 missed wales is ten loop pitches. Obviously, the average loop length of an elastomeric yarn in the fabric with 2 missed wales at the knitting repeat is the highest and the average loop length of an elastomeric yarn in the fabric with 10 missed wales at the knitting repeat is the smallest. It should be noted that the difference in the lengths of elastomeric yarn of the boundary options of missed knit repeat does not exceed 6 %.

Analysis of the results showed the greatest impact the miss knit repeat has on the number of wales (N_w) per 100 mm (Fig.3.a) and the thickness (T) of fabric (Fig.3.b). The number of wales is increased by 14 % and the thickness is reduced by 10 % with increasing the number of missed wales from 2 to 10.

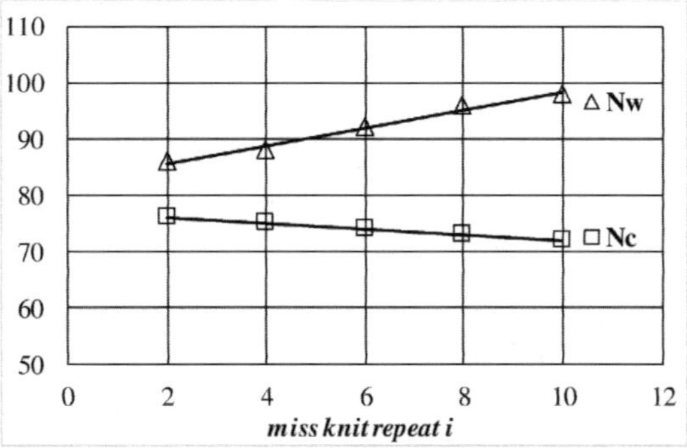

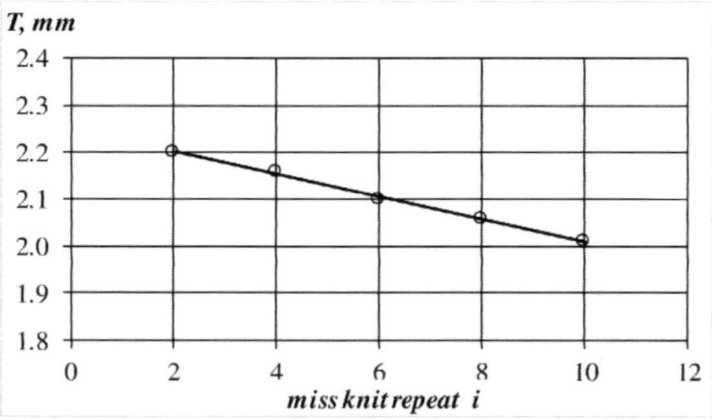

Figure 3. Stitch density (a) and thickness (b) of elasticized fabric

The thickness of the fabric with weft-inserted yarn is determined by the its diameter. Changes in the thickness of a knitted fabric can be explained by the different positions of elastomeric yarns in the knitted structure. According to the interlooping repeat (Fig. 2.d) two elastomeric yarns are laid in the course (Fig. 4). The loops from the ground yarn wrap them on both sides of fabric. This ground loops form curved configuration in the knitted structure. The two elastomeric yarns are positioning beside one another in the knitted structure with 2 missed wales at the knitting repeat (Fig. 4.a). The thickness of such structure equals two diameters of elastomeric yarns and two diameters of ground yarns. The two elastomeric yarns are positioning one over another in the knitted structure with 10 missed wales at the knitting repeat (Fig. 4.b). The thickness of this structure equals one diameter of elastomeric yarns and two diameters of ground yarns. Moreover, in this structure the ground loops are less curved and more extended walewise (Fig. 4.b). This leads to an increasing of the loop height and correspondingly to reducing up to 5% the number (Nc) of courses per 100 mm (Fig.3.a).

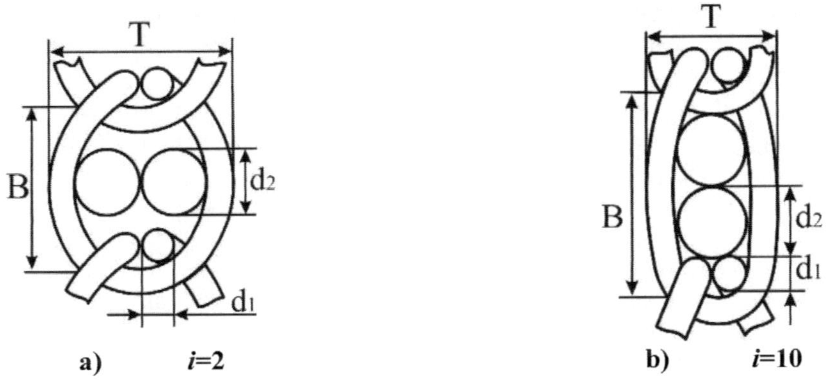

Figure 4. The positioning of filling yarns in knitted structure

The number of missed wales in knitting repeat does not affect the basis weight of elasticized knitted fabric. The difference in basis weight of the boundary options of miss knit repeat does not exceed 3%.

4.2. Mechanical properties

Extensibility and elasticity are the main characteristics of the elasticized fabric that determine their use for medical purpose. The experimental results show that full stretch of the knitted fabric with two missed wales at repeat is less than 75 % and does not satisfy the requirements. Full stretch of the knitted fabric increases with increase of number of missed wales and reaches 150 %. Different values of the full stretch are associated primarily with varying degrees of relaxation of elastomeric yarn in knitted structure of different variants.

As shown on the plots (Fig. 5) the elastic stretch is the largest part of the full stretch of elasticized knitted fabric. Quota of elastic part is more than 0.92 (Table 1), which is associated with the use of elastomeric yarn. Elastic stretch of the knitted fabric increases with increase of number of missed wales too as full stretch and depends on degrees of relaxation of elastomeric yarn in knitted structure.

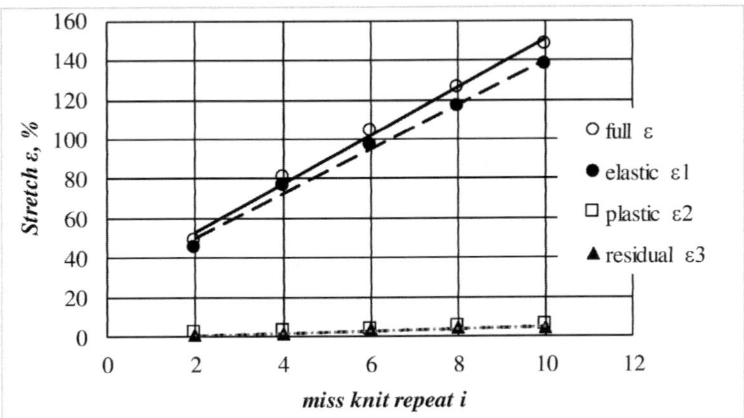

Figure 5. Full stretch and its parts

As for the plastic and residual components of the full stretch, they also increase with increase of number of missed wales at repeat (Fig. 5), but the maximum absolute value is respectively $\varepsilon_2 = 6{,}2$ % and $\varepsilon_3 = 4{,}6$ %. Thus, elasticized knitted fabric of different miss knit repeat can show acceptable dimensional stability.

5. Conclusions

The experimental data showed that the missed knit repeat affects the relaxation degree and position of an elastomeric yarn in the knitted structure, which in turn leads to changes in the structure parameters and elasticity of knitted fabrics.

The elasticized fabric with 4 ÷ 8 missed wales between the loops from the elastomeric yarn is recommended for using in abdominal binders in order to provide the necessary extensibility (80 ÷ 120 %) with low residual strain (2 ÷ 4%).

References

[1] Oks BS, Oks AB 2008 Compressive knitted items for treatment of chronoc venous insufficiency and lymphostasis. *44 Congress IFKT* "Knitting round the clock" (Saint-Petersburg, Russia)

[2] Bulent Cakmak et al. 2014 The Effect of Maternity Support Belts on Postural Balance in Pregnancy, *The American Academy of Physical Medicine and Rehabilitation* **6**, 624-628

[3] Melnyk LM, Kyzymchuk OP 2012 The utilization of elastic knitted materials in the medical-preventive goods. *Bulletin of KNUTD* **4**, 139-145 – in Ukranian

[4] Polyakova S, Truevtsev A, Stolyarov O 2008 Tensile and stress relaxation behavior of the highstretch weft-knitted fabrics containing elastomeric yarns. *44 Congress IFKT* "Knitting round the clock" (Saint-Petersburg, Russia)

[5] Hui CL and Ng SF 2003 Theoretical Analysis of Tension and Pressure Decay of a Tubular Elastic Fabric, *Textile Research Journal* **73**, 268-272

[6] Senthilkumar M and N Anbumani N 2011 Dynamics of Elastic Knitted Fabrics for Sports Wear, *Journal of Industrial Textiles* **41**, 13

[7] Melnyk L, Omelchenko V 2008 Technology of weft knitted elastic fabric with an additional fastening of filling thread, *Bulletin of KNUTD* **1**, 62-65. – in Ukranian

[8] Abdessalem S et al. 2009 Influence of Elastane Consumption on Plated Plain Knitted Fabric Characteristics, *Journal of Engineered Fibers and Fabrics*, **4**, 30-35

[9] Mishta V 1991 Designing the structural parameters and properties of elastic weft knitted fabric. PhD manuscript: 05.19.03. Kyiv, 174 p. – in Russian

[10] Sadek R et al. 2012 Effect of Lycra Extension Percent on Single Jersey Knitted Fabric Properties. *Journal of Engineered Fibers and Fabrics* **7**, 11-16

[11] Kyzymchuk O and Melnyk L 2013 The using of reinforced elastomeric yarn of GUMEX on flat knitting machine, *Technologies & Design* **2** – in Ukrainian

[12] Melnyk L and Kyzymchuk O 2013 Effect of knitting method of elastomeric threads on the structure parameters and properties of knit, *International Scientific Conference UNITECH* (Gabrovo, Bulgaria), 207-211 – in Russian

IOP Conf. Series: Materials Science and Engineering **141** (2016) 012007 doi:10.1088/1757-899X/141/1/012007

Non-destructive X-ray examination of weft knitted wire structures

M Obermann[1], M Ellouz[2], S Aumann[1], Y Martens[2], P Bartelt[2], M Klöcker[2], T Kordisch[2], A Ehrmann[2,3] and M O Weber[1]
[1] Niederrhein University of Applied Sciences, Faculty of Textile and Clothing Technology, Mönchengladbach, Germany
[2] Bielefeld University of Applied Sciences, Faculty of Engineering and Mathematics, Bielefeld, Germany

E-mail: andrea.ehrmann@fh-bielefeld.de

Abstract. Conductive yarns or wires are often integrated in smart textiles to enable data or energy transmission. In woven fabrics, these conductive parts are fixed at defined positions and thus protected from external loads. Knitted fabrics, however, have relatively loose structures, resulting in higher impacts of possible mechanical forces on the individual yarns. Hence, metallic wires with smaller diameters in particular are prone to break when integrated in knitted fabrics. In a recent project, wires of various materials including copper, silver and nickel with diameters varying between 0.05 mm and 0.23 mm were knitted in combination with textile yarns. Hand flat knitting machines of appropriate gauges were used to produce different structures. On these samples, non-destructive examinations, using an industrial X-ray system Seifert x|cube (225 kV) equipped with a minifocus X-ray tube, were carried out, directly after knitting as well as after different mechanical treatments (tensile, burst, and washing tests). In this way, structural changes of the stitch geometry could be visualized before failure. In this paper, the loop geometries in the knitted fabrics are depicted depending on knitted structures, wire properties and the applied mechanical load. Consequently, it is shown which metallic wires and yarns are most suitable to be integrated into knitted smart textiles.

1. Introduction

Smart textiles contain sensory or actuatory elements which can react to and interact with the environment. Such elements can, e.g., be used to measure breathing frequency, ECG or pulse [1], skin temperature, sweating and other vital signs. They can illuminate clothing for safety or design purposes, e.g. by LEDs or optical fibers [2]. Reactions on temperature changes and other environmental conditions can also be performed by textile-integrated phase change materials or similar "intelligent" materials [3].

In many smart textiles, electronic components play an important role to achieve desired functionalities. These parts have to be connected to transport energy and data. Conductive paths in textiles can either be generated by common wires or by conductive yarns, with advantages and problems occurring in both cases. While wires tend to work hardening and breaking during washing and use, conductive yarns may also break or oxidize, depending on their composition.

[3] To whom any correspondence should be addressed.

Especially the integration of conductive yarns into textiles has been examined in recent years. Several techniques, such as weaving, sewing and embroidery, were used [4,5]. Connection technologies between non-textile sensors and actuators and textile panels [6] or yarns [7] were investigated, too. These projects, however, dealt with non-elastic textiles as bases for conductive paths.

Since neither wires nor conductive yarns are really elastic – even so-called "elastic" conductive yarns can only be elongated by a few percent [8, 9] –, both solutions necessitate nonlinear integration in sensory underwear and other elastic clothing and textiles. It has been shown that sewing conductive yarns in undulating forms meets the demands of elastic underwear [10].

A simpler method, however, is to directly use conductive yarns or wires during knitting to produce an elastic fabric with integrated conductive paths. This approach was followed in a recent project, aiming at a comparison between wires of different materials and diameters. Additionally, one conductive yarn was included in the experiment. An overview of the samples under examination is given in the next section.

2. Experimental

Table 1 depicts the different wires and an additional conductive yarn which were used in the recent study.

Table 1. Wires and conductive yarn under examination.

Material	Diameter
Cu (copper)	0.23 mm
CuAg7 (copper silver 7 %)	0.15 mm
Ni (nickel)	0.10 mm
NiCr (nickel-chrome)	0.10 mm
StSt (stainless steel) AISI 316 L	0.05 mm
Shieldex (silver-plated nylon 66), prod. by Statex	235/34 dtex 2-ply

For knitting, a single-bed hand flat knitting machine SilverReed SK-280 with gauge E5.5 was used. The following structures were produced: single jersey (all materials); stitches and tucks alternating in course and wale direction (samples T), and stitches and floats alternating in course and wale direction (samples F) (NiCr only). All samples included 5 single courses of conductive material, separated by 8-10 courses of non-conductive yarn (PAN Nm 22/2).

The samples underwent the following tests: washing at 60 °C with heavy-duty detergent and concurrent spin-cycle of 800 min^{-1}; bursting tests using a TruBurst 611 (James H. Heal) with 22 mm bulging height for 10 cycles; and elongation tests using a tensile testing machine with 25 % maximum elongation for 10 cycles. In the latter test, the samples were elongated along the courses.

The samples were examined using the industrial X-ray system X-Cube 225 of GE Measurement & Control Solutions with a minifocus after knitting as well as after the aforementioned tests. The X-ray parameters are: 64 kV, 2.1 mA, focus diameter 0.4 mm, exposure time 100 ms, gain 1100, 16 iterations, and flash filter. In this way, the conductive courses can be depicted separately, without the overlaps by non-conductive yarns which would occur in an optical microscope and hide parts of the conductive materials.

3. Results

During knitting, the following experiences were made: The Cu wire needed to be unwound tangentially because the twist in the wire became too strong to guide it correctly through brake, thread guides and yarn carrier. It never broke but sometimes caused missed stitches.

CuAg7 was even harder to knit due to its high bending stiffness. During the 4x5 courses, it broke 3 times.

Ni and NiCr were easily unwound and knitted; they never broke.

Stainless steel had to be knitted slowly and tended to get in contact with the non-conductive yarn alongside. Due to its elastic behavior (opposite to Cu which tends to bending plastically), it was hard to keep away from the other yarn. It broke 5 times during the 4x5 courses.

Shieldex yarn, finally, behaves identical to each "usual" yarn and is knittable very well.

Samples of all materials and structures were examined using X-ray directly after knitting. Fig. 1 shows the results (only one representative conductive line per sample).

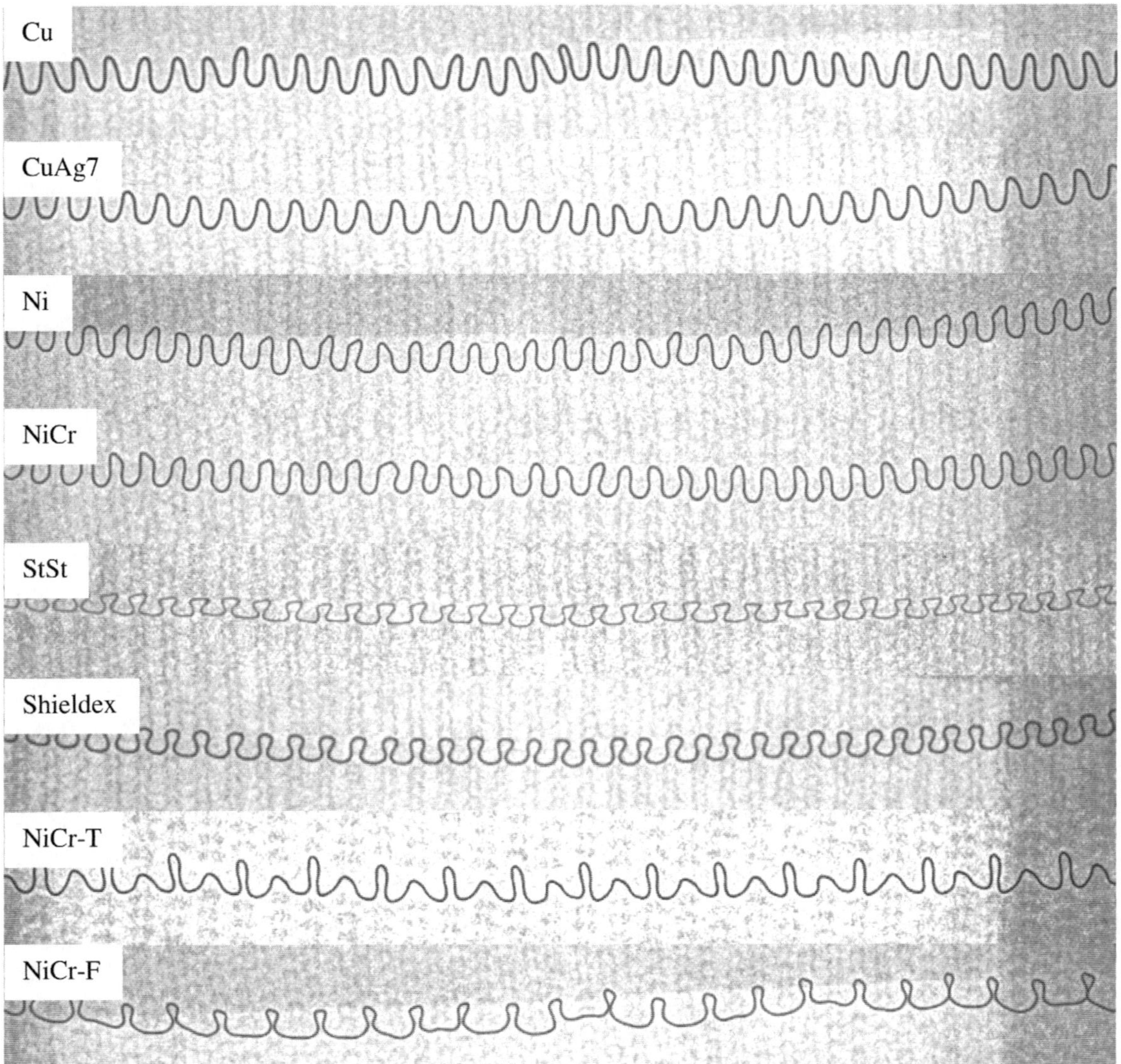

Figure 1. X-ray images of all samples directly after knitting. Shadows can be attributed to the sample holder. Sample names are explained in Table 1 and the text below.

Apparently the Cu wire is too thick to create stitches in the usual shape. Irregularities are visible in all knitted courses (here in the middle of the course, with bent stitches and decreased loop widths).

CuAg7 looks much more regular, with the stitches all bent in the same direction, as usual for typical yarns.

Ni and NiCr show similar shapes, with the loop heads being more pronounced than in CuAg7, but some irregularities.

Knitting with stainless steel results in over-pronounced loop heads and strong buckles in the stitch shapes.

Shieldex produces common stitch shapes, as expected from a usual yarn.

Both additional knitted structures exhibit the expected loop shapes with not completely rounded stitch heads and some irregularities in the complete courses.

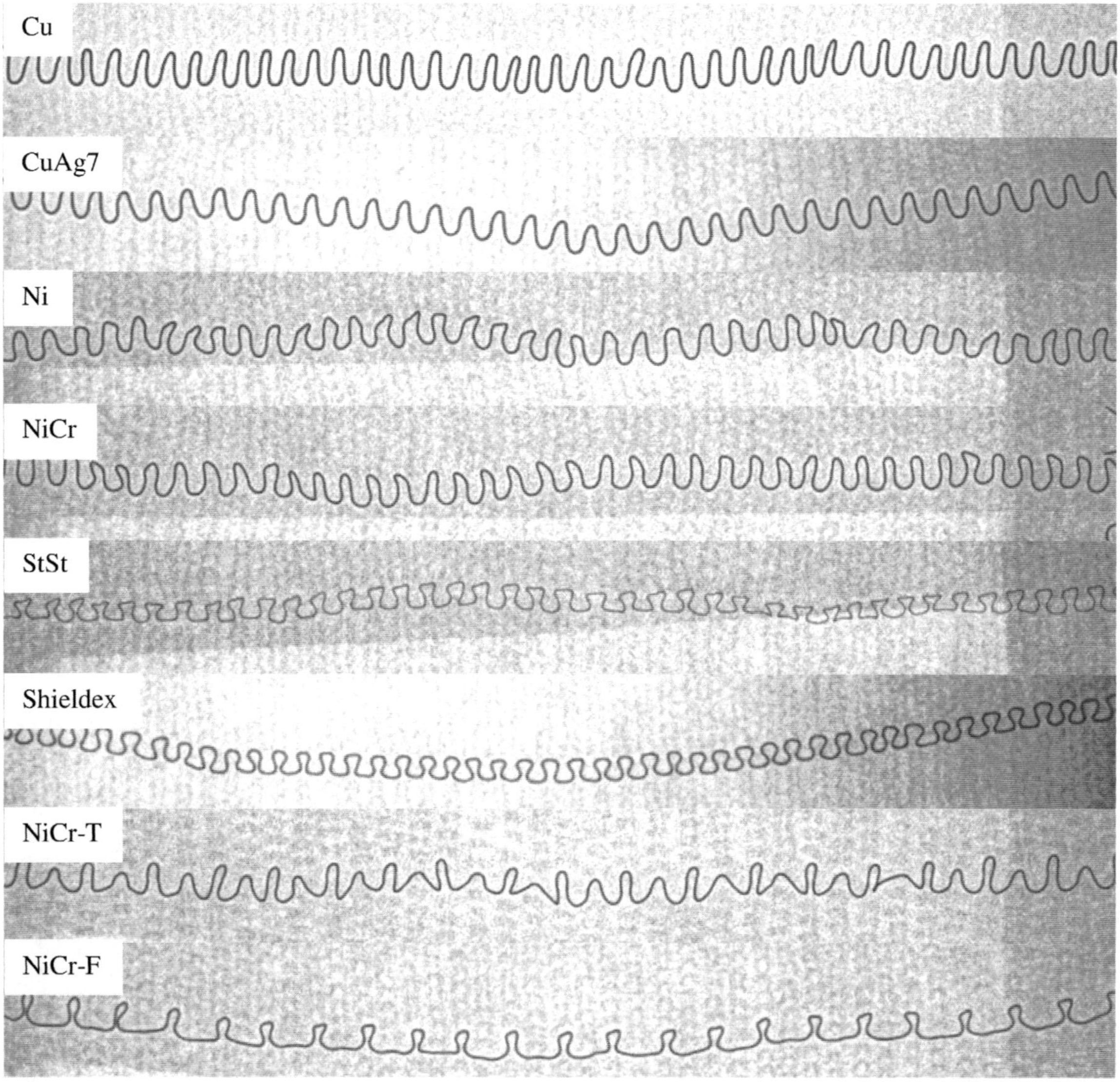

Figure 2. X-ray images of all samples after washing. Sample names are explained in Table 1 and the text below.

Fig. 2 shows the samples after washing. Cu and CuAg7 stitches do not change their shapes visibly. NiCr and especially Ni, however, show much more irregular structures after washing. The same observation applies to the stainless steel stitches. Shieldex – as expected for a nylon yarn with only thin metallic coating – retains its regular structure. Regarding the additional knitted structures from NiCr, the T version looks more irregular than directly after knitting, while the F version shows only small deviations from the original shape.

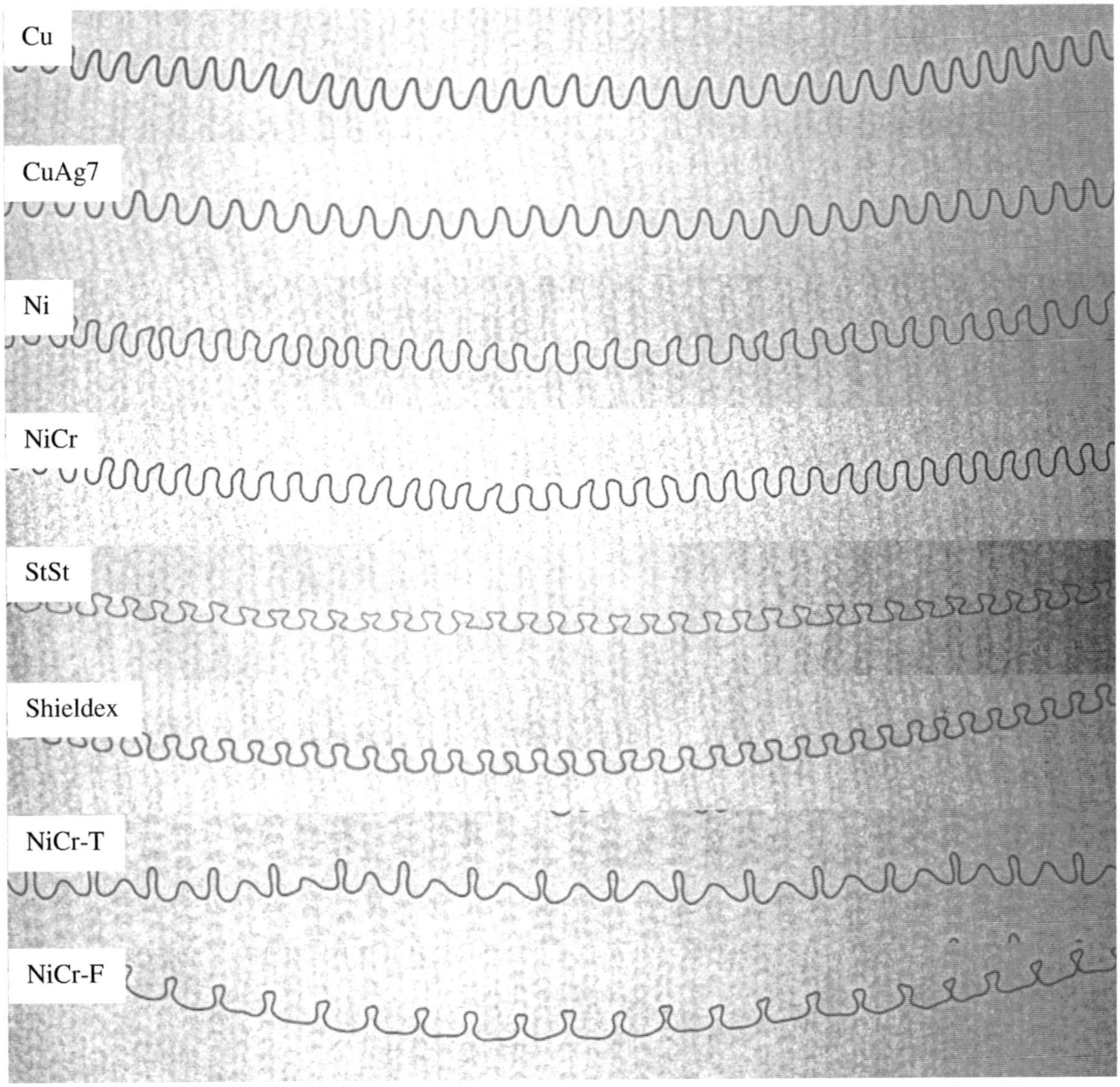

Figure 3. X-ray images of all samples after burst tests. Sample names are explained in Table 1 and the text below.

In Fig. 3, X-ray images of the samples after repeated burst tests are depicted. Cu and CuAg7 again do not show significant deviations from their original shape. Ni and NiCr also show less deformation than after washing. The same finding holds for the other materials. Apparently, the regular tension in the repeated burst tests is less strenuous for the wires than the irregular stress during washing.

The results of repeated linear elongation by 25 % are depicted in Fig. 4. Similarly to the results of the aforementioned demands, Cu and CuAg7 are nearly unaltered in comparison to the original knitted structures. The nickel wire, however, has significantly changed its form. Opposite to the bursting test in which an isotropic force is applied on the knitted fabric, in the tensile test only a force along the courses works on the textile. This leads to a significant difference between the shape of the loops before and after elongation.

NiCr and stainless steel wires remain apparently unaltered. In the same way, NiCr in T and F version stay nearly unchanged.

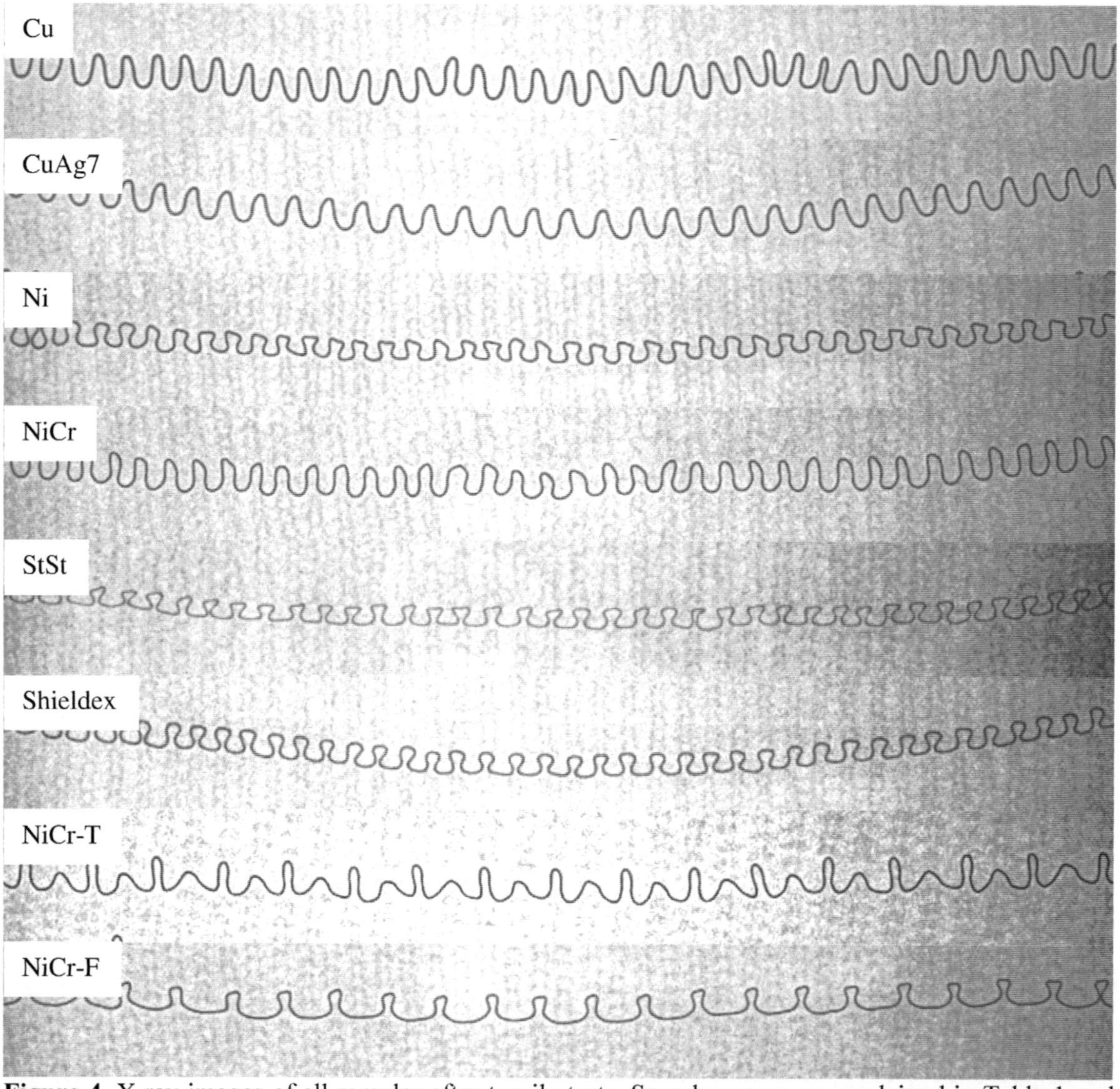

Figure 4. X-ray images of all samples after tensile tests. Sample names are explained in Table 1 and the text below.

Another important parameter of conductive materials, besides their mechanical properties, is their conductivity. Thus, the resistances of all conductive materials under examination were measured directly after knitting as well as after washing, bursting and elongation tests, using a multimeter with clamps. It should be mentioned that in this way, the overall resistance is composed of the contact resistance between multimeter clamps and knitted wire / yarn, and the linear resistance in the conductive material.

In all wires, the overall resistance stayed unaltered after washing, bursting and elongation tests. A significant difference, however, was found for Shieldex yarn after washing, where the resistance increased by approximately a factor of 2-2.5. Since microscopic inspection of the Shieldex yarn after washing did not reveal a possible reason for this finding, this issue will be investigated further in the near future.

Table 2. Rating of the materials under investigation during knitting and subsequent tests, with positive (+), negative (-) and neutral (0) rating.

Material	Knitting	Washing	Bursting	Elongation
Cu	-	+	+	+
CuAg7	-	+	+	+
Ni	0	-	0	-
NiCr	0	-	0	0
Stainless steel	-	-	0	0
Shieldex	+	-	+	+

To conclude, Shieldex yarn is ideally suited for applications which do not need to be washed. In other applications, wires may be good alternatives, depending on the requirements and possibilities during knitting as well as the mechanical demands during usage. Especially Cu and CuAg7 show good resistance against washing, bursting and elongation. Other materials and wire diameters may be more suitable for other applications.

Nevertheless it must be stated that none of the materials under examination can be attributed as perfect for all applications. This finding shows the necessity of further investigations, taking into account other conductive yarns, stranded wires etc.

4. Conclusion

Conductive yarns or wires, used for energy or data transmission in smart textiles, still suffer from some issues, compared to usual textile materials. They may be hard to knit and change their mechanical and electrical properties during washing or mechanical stress.

In a recent project, we have shown the possibilities of investigating conductive yarns and wires integrated in textile fabrics by X-ray examinations. In this way, the advantages and disadvantages of different wires and a conductive yarn could be shown.

In future projects, different wires and conductive yarns will be examined further, using the X-ray technology to enable visual inspection of the complete conductive materials' shapes without impairment by the neighboring non-conductive courses.

5. References

[1] Stoppa M and Chiolerio A 2014 Wearable Electronics and Smart Textiles: A Critical Review, *Sensors* **14** 11957-92
[2] Schwarz-Pfeiffer A, Mecnika V, Beckers M, Gries T, and Jockenhoevel S 2015 *Optical Fibers*, in: Tao X, editor. Handbook of Smart Textiles. Singapore: Springer, p. 79-108
[3] Mondal S 2008 Phase change materials for smart textiles – An overview, *Appl. Thermal Eng.* **28** 1536-50
[4] Haase E, Pusch T, Ezzeddine L, Cherif C, Illing-Günther H, Zschenderlein D, Möhring U, Fritzsche B, Haiduk F, Nowottne H.-J. 2006 Energie- und Informationsübertragung in Smart Textiles, *Technische Textilien* **49** 256
[5] Institut für Textil- und Bekleidungstechnik of TU Dresden. Energie- und Informationsübertragung in Smart Textiles, project description, https://tu-

dresden.de/die_tu_dresden/fakultaeten/fakultaet_maschinenwesen/itm/forschung/forschungsthemen/smart_textiles/index_html (accessed 11-02-2016)

[6] Kotterba B, Dömer B, Meier H, Alich I, Jäger M, Horter H, Linti C 2008 TEMONICS – Telemonitoring with cardiorespiratory systems, *Proceedings of Ambient Assisted Living – AAL – 1. Deutscher AAL-Kongress*

[7] Tillmanns A, Heimlich F, Brücken A, et al. 2008 Investigation of the possibilities to realize light effects in textile fabrics, *Technical Textiles* **4** E182–E183

[8] Ohmatex. Brochure, http://www.ohmatex.dk/pdfer/Ohmatex_brochure.pdf (accessed 11-02-2016)

[9] Novonic® Die beste Verbindung von Textil, Elektronik und Zukunft, http://www.novonic.de/web/zimsi.nsf/id/pa_novonic.html

[10] Aumann A, Trummer S, Brücken A, Ehrmann A and Büsgen A 2014 Conceptual design of a sensory shirt for fire-fighters, *Text. Res. J.* **84** 1661-1665

IOP Conf. Series: Materials Science and Engineering **141** (2016) 012008 doi:10.1088/1757-899X/141/1/012008

Smarten up garments through knitting

A Schwarz-Pfeiffer[1], M Obermann[1], M O Weber[1] and A Ehrmann[2]
[1] Niederrhein University of Applied Sciences, Faculty of Textile and Clothing Technology, Mönchengladbach, Germany
[2] Bielefeld University of Applied Sciences, Faculty of Mathematics and Engineering Sciences, Bielefeld, Germany

E-mail: anne.schwarz-pfeiffer@hsnr.de

Abstract. Smart textiles are a growing and fascinating field with enormous potential in the field of wearable electronics: shirts with integrated electrodes, socks stimulating the blood circulation or heating clothing are just a few examples of wearable, smart textile products. Most often, the technology of choice for on-the-body-worn smart textiles is knitting as it results in stretchable and, hence comfortable garments. This presentation explores the knitting technology in respect to smart textiles giving an overview of current research activities as well as commercially available products on the market. It further intends to foster the transfer of research approaches into business applications as well as to develop new challenging research ideas.

1. Introduction

We are living in a knowledge driven society that is facing an increasing impact of science and technology on all aspects of life through products and services, and consumer needs. More and more objects surrounding us in our everyday life get smart and so textiles do as well. The field of smart textiles is not a discrete area; it is more an interdisciplinary subject incorporating science, technology, design and human sciences, and its future lies in the potential of combining different technologies. When talking about smart textiles we often talk about the convergence of textiles and electronics – a combination of materials capable of accomplishing a wide spectrum of functions to increase, for instance, our well-being and health [1]. Smart functions, however, can also be integrated using phase change materials [2], optical fibers [3], etc. Such smart textiles can be used in a broad range of applications, from automotive to civil engineering to sports and medical purposes.

A smart textile textile material interacts actively with its environment, i.e. it responds or adapts to changes in the environment or to an external input. To be able to carry out this interaction, it is often incorporated into a system, comprising textile and non-textile parts. The system ideally comprises six different components: sensor, actuator, internal and external communication, power source and data processor [4]. Each of these components has been realized at least partly out of woven, knitted, braided fabrics or non-wovens over the past years [5,6] and research activities are still constantly increasing. Against this background, this paper gives an overview on recent advances in the research and development of these components. Above that, as the research diversified over the past years, we concentrate on knitted fabrics and highlight their uses to realize different smart textile components.

2. Knitted sensors and electrodes

One of the typical properties of knitted fabrics is their stretchability. In a knitted fabric made of conductive yarn, the resistance will change during elongation [7]. Thus, knitted fabrics can be used as stretch sensors, with the exact properties dependent on yarn and knitted structure (Fig. 1).

Farringdon et al., e.g., have examined knitted strips and found a linear increase over a broad range of elongations [8]. Such a sensor can, e.g., be used to measure arm or leg movements.

Large strains occurring at high temperatures can similarly be measured using knitted fabrics from stainless steel or carbon fibers, with the latter showing higher sensitivity and better repeatability [9].

The elongation-dependent resistance can also be used to detect breathing frequency. Using electro-conductive yarns, Zieba and Frydrysiak have produced knitted samples with conductive stripes from polyacrylonitrile staple fibers coated with copper sulphide (produced by Euro-static), combined with non-conductive parts from cotton and polyester. Elongation in vertical and horizontal direction have shown to produce differences in the measured resistance, allowing for using such samples as breathing sensors [10]. Combinations of stainless steel yarn and elastane parts have also been shown to be suitable for measuring the respiration by resistance and inductance changes [11]. Nevertheless the dependence of the sensing properties on yarn type, yarn input tension, knitted structure and other parameters must be taken into account [12].

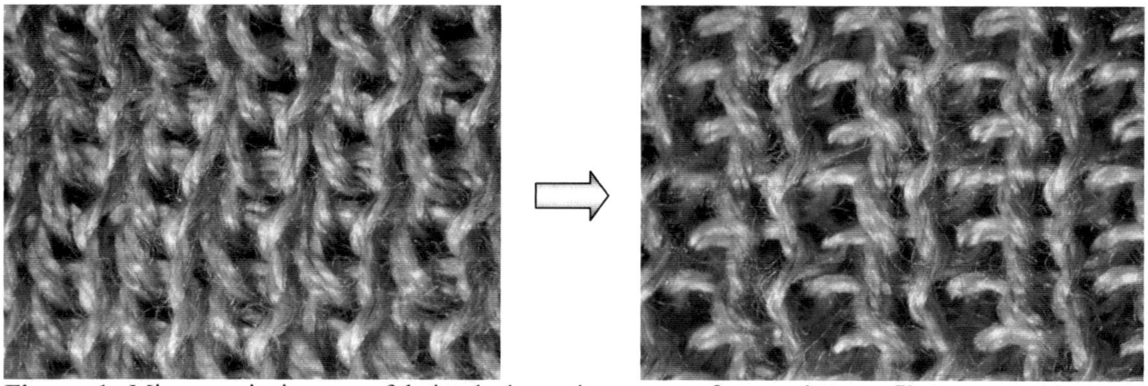

Figure 1. Microscopic images of knitted elongation sensor from polyester fiber yarn with 20 % stainless steel fibers in relaxed (left panel) and stretched state (right panel).

Another possibility to measure respiration is using piezoelectric crystals which generate electric voltage due to deformation during breathing [13]. Piezoresistive sensors can be produced by coating conductive polymer sensors on knitted or woven fabrics, often resulting in higher gauge factors (changes in resistance per strain) than in conventional elongation sensors [14]. Such sensors can also be used to create gloves or complete wearable motion capture systems for man-machine interactions [15,16].

Finally, breathing can be measured using optical fibers [10]. The static deformation of a knitted sample with included optical fibers leads to a change of the light which is led through the optical fiber onto a photodiode. The photodiode voltage is approximately linearly correlated with the elongation. Such macro-bending transducers can be easily achieved by integrating optical fibers in half-loop or U shape. Stretching the fabric results in an increased curvature radius and thus reduced losses in the fiber [17]. Signal analysis in single-mode optical fibers, however, is impeded by mode coupling between the fundamental mode and reflection modes, resulting in undesired oscillations in the curves of loss due to fiber bending [18-20]. This problem has to be coped with using defined wavelengths [17]. Alternatively, multimode fibers can be used in which the special intensity of the modal power is modulated; the change of the intensity profile can be related to the stress applied on the fiber [21].

IOP Conf. Series: Materials Science and Engineering **141** (2016) 012008 doi:10.1088/1757-899X/141/1/012008

Conductive knitted fabrics do not only respond to elongation but also to pressure. Especially spacer fabrics with two conductive sides, separated by a non-conductive spacer thread (Fig. 2), can be used as textile switches or pressure sensors [22]. It is not only possible to measure their resistance but also to use them as parallel-plate capacitors with a capacity inversely proportional to the distance between the plates. A spacer fabric in the shape of an array of parallel plate capacitors enables to localize pressure on the fabric and in this way, e.g., designing a sensor which can measure sitting postures or lying orientations [23].

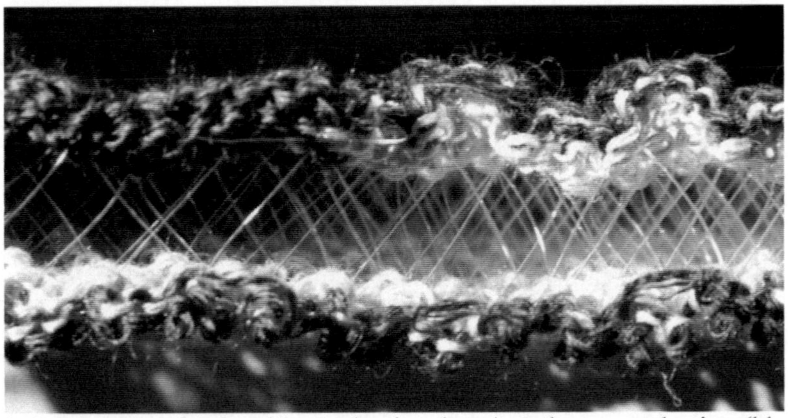

Figure 2. Weft-knitted spacer fabric with conductive (grey) and non-conductive (blue) yarns in both planes, separated by a non-conductive spacer thread.

Knitted electrodes can also be used as ECG or pulse sensors [24,25]. Additional to the usual problems of textile ECG electrodes, i.e. varying resistances between skin and electrodes as well as motion artifacts, knitted electrodes have the additional disadvantage of varying resistances inside the electrode due to elongation effects [26,27]. Nevertheless, knitted ECG sensors have been developed by several research groups in different projects, such as MyHeart [28-31] or MERMOTH [32,33]. The ECG signals can be increased by higher contact pressures and additional moisture [34]. However, in many cases more stable and reliable ECG signals can be detected by non-elastic woven electrodes which are integrated in a knitted sensory shirt [35].

Finally, knitted fabrics can also be used as a base for a sensory coating, detecting chemical elements such as ammonia [36].

3. Knitted actuators

Actuators transform electrical or other signals into outputs which can be recognized by people, e.g. in form of light, sound, or motion. Such signals can be used to inform the wearer of a smart garment about the input gained by textile or textile-integrated sensors, can warn them if any threshold values are exceeded, or just react on a detected change in the environment by counteracting, e.g. by heating up a textile if the skin temperature of the user gets too low.

One of the possibilities to include a tactile actuator is based on shape memory materials, such as shape memory alloys (SMAs) or shape memory polymers (SMPs). Such materials can consist of chemically crosslinked glassy thermosets or semi-crystalline rubbers as well as of physically crosslinked thermoplastics or block copolymers [37]. In all cases, they react on temperature, resulting in thermomechanical deformation and recovery [38,39].

They can be used as actuators as well as sensors. First tests were performed with respect to integration of such materials in knitted fabrics [40]. In this way, interesting shapes and time-dependent effects can be added to knitted fabrics, while the manufacturing process is still challenging [41].

The integration of light in textiles can be performed using LEDs, optical fibers, or electroluminescent (EL) wires. While LEDs are usually integrated in textile fabrics by sewing, embroidery or gluing [42], optical fibers can be knitted (see examples in chapter 2). EL-wires can easily be included in woven fabrics; knitting them, however, can easily lead to the destruction of the electroluminescent effect. Additionally, it should be mentioned that they require a high voltage supply which makes them less easily usable in clothing than low-voltage LEDs [43]. Nevertheless, several research groups have developed methods to create electroluminescent coatings on textile materials, mostly on woven fabrics which are not as flexible as knitted fabrics [44-46].

As a last example for knitted actuators, heating elements have to be mentioned. Heating clothing is useful especially for people working outside. Heated knitted fabrics can be composed of conductive yarns, e.g. from metal or metal coated filaments in combination with textile filaments, which are heated due to resistive losses if a current is flowing through the fabric [47]. The temperature reached in this way depends on the parameters of the power source as well as on those of the yarn, especially its resistivity and cross section. By introducing a control unit, the heated clothing can be regulated with respect to external and skin temperatures [48,49]. This is of special importance since not only the manufacturing method but also the contact pressure at the binding points influences the degree of heating [50].

4. Knitted transmission lines

Signal and energy transmission in smart textiles necessitates conductive fibers or yarns. Depending on the desired application, especially on the distinction between direct current and alternating current in the high frequency range, and on the textile production method, different wires or yarns have to be chosen. Additionally, crosstalk between neighboring transmission lines has to be taken into account since conductive fibers and yarns are usually not electrically shielded [51].

Conductive threads can, e.g., be created by filling fibers with conductive materials such as carbon or metal powder, coating of fibers with metals or conductive polymers, or using filaments and fibers which are inherently conductive, i.e. from metal or conductive polymers [52]. Fig. 3 shows two of these alternatives. Usually, compromises between knittability, washability, conductivity and other desired parameters are necessary.

Transmission lines can either directly be implemented or afterwards be added using printing, coating or spraying electroconductive materials as well as embroidering or sewing [53]. Integration of conductive materials in knitted fabrics can be used to produce antennas, e.g. for RFID or bluetooth applications [54].

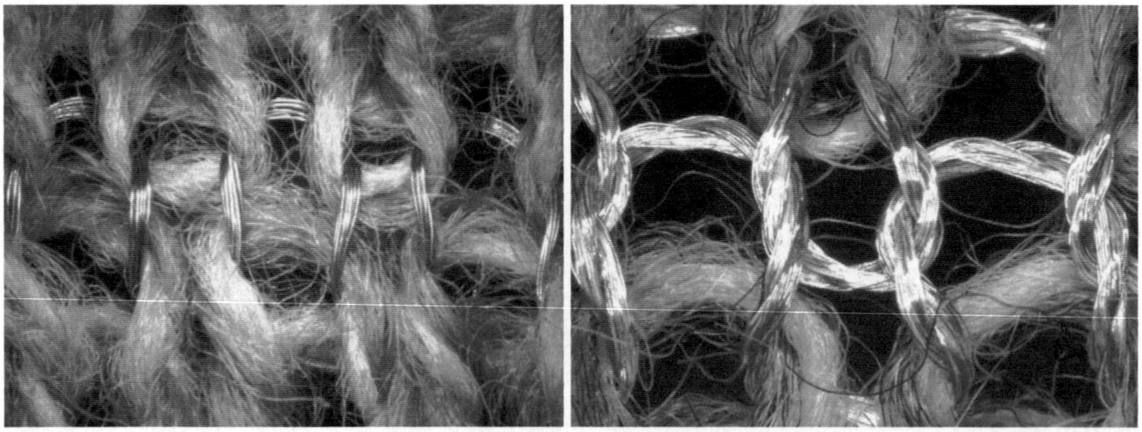

Figure 3. Conductive transmission lines from copper strand (left panel) and silver-coated polyamide filaments (right panel) in single-jersey knitted fabrics from non-conductive materials.

IOP Conf. Series: Materials Science and Engineering **141** (2016) 012008 doi:10.1088/1757-899X/141/1/012008

5. Other knitted smart textile components

Smart textiles using electronic elements need power supplies. To develop purely textile power sources, several attempts have been made. Energy storing technologies can be separated in three different approaches: textile capacitors, textile pseudo-capacitors, and textile batteries [55]. Besides coating layer systems on textile fabrics [56], it is possible to create fiber-capacitors and -batteries and to create woven or knitted textile fabrics from them [57-59]. Nevertheless, problems may occur especially during the knitting process which is challenging for the functionalized yarns [60]. This may lead to damages of the coating and resultant loss of the desired energy storage capacity.

In future development, such mechanical impacts have to be taken into account as well as the necessary increase of energy storage capacity to allow for long-time use of smart textiles without non-textile parts.

6. Challenges and opportunities

Knitted fabrics are of favorable use among the desired fabric constructions when it comes to on body worn electronics due to their stretchability and hence comfort properties. However, as mentioned in this paper, elasticity is often a property that counteracts with reliable electronic properties.

Hence, a challenge is the development of knitted structures with different elasticity levels throughout the fabric. In this way, the conductive or functional parts can be stiff, while the nonfunctional part is highly stretchable.

Another challenge has to be addressed from the material side. Yet, many interesting functional materials, like piezoelectric, chromic materials or hydrogels changing volumes, exist, but in a form that is not processable through knitting. Here, advances in functional and smart fibers and yarns have to be made.

The integration and interconnects are also a widely discussed drawback in the field of smart textiles. This challenge will only be overcome if knitting machineries adapt to process novel materials as well. An example of good practice can be found in the embroidery sector, in which machines can already process LEDs and connect them with embroidered conductive yarn traces fully automatically.

Last but not least, once knitted smart textiles are ready to enter the market, standards are needed for documentation of the performance and safety of these new products by setting up new test procedures. A step towards this direction is done by the European Committee for Standardization (CEN) to create a working group around smart textiles. Hopefully, more initiatives in this vein will be started in the near future to enhance the smart textile market.

7. References

[1] Stoppa M and Chiolerio A 2014 Wearable Electronics and Smart Textiles: A Critical Review, *Sensors* **14** 11957-92

[2] Mondal S 2008 Phase change materials for smart textiles – An overview, *Appl. Thermal Eng.* **28** 1536-50

[3] Schwarz-Pfeiffer A, Mecnika V, Beckers M, Gries T, and Jockenhoevel S 2015 *Optical Fibers*, in: Tao X, editor. Handbook of Smart Textiles. Singapore: Springer, p. 79-108

[4] Schwarz A, van Langenhove L, Guermonprez P, and Deguillemont D 2010 A roadmap on smart textiles, *Textile Progress* **42** 99-180

[5] Haase E, Pusch T, Ezzeddine L, Cherif C, Illing-Günther H, Zschenderlein D, Möhring U, Fritzsche B, Haiduk F, Nowottne H.-J. 2006 Energie- und Informationsübertragung in Smart Textiles, *Technische Textilien* **49** 256

[6] Institut für Textil- und Bekleidungstechnik of TU Dresden. Energie- und Informationsübertragung in Smart Textiles, project description, https://tu-

dresden.de/die_tu_dresden/fakultaeten/fakultaet_maschinenwesen/itm/forschung/forschungsthemen/smart_textiles/index_html (accessed 11-02-2016)

[7] Ehrmann A, Heimlich F, Brücken A, Weber M O, Haug R 2014 Suitability of knitted fabrics as elongation sensors subject to structure, stitch dimension and elongation direction, *Text. Res. J.* **84** 2006-2012

[8] Farringdon J, Moore A J, Tilbury N, Church J and Biemond P D 1999 Wearable sensor badge and sensor jacket for context awareness, *The Third International Symposium on Wearable Computers* 107-113

[9] Zhang H, Tao X, Yu T and Wang S 2006 Conductive knitted fabric as large-strain gauge under high temperature, *Sensors and Actuators A: Physical* **126** 129-140

[10] Zieba J, Frydrysiak M 2006 Textronics – Electrical and Electronic Textiles. Sensors for Breathing Frequency Measurement, *Fibers and Textiles in Eastern Europe* **14** 43-48

[11] Catrysse M, Puers R, Hertleer C, van Langenhove L, van Egmond H and Matthys D 2004 Towards the integration of textile sensors in a wireless monitoring suit, *Sensors and Actuators A* **114** 302-311

[12] Atalay O and Kennon W R 2014 Knitted Strain Sensors: Impact of Design Parameters on Sensing Properties, *Sensors* **14** 4712-4730

[13] Paradiso R, Loriga G, Taccini N 2005 A wearable health care system based on knitted integrated sensors, *IEEE Transaction on Information Technology in Biomedicine* **9** 337-344

[14] Calver T, Patra P, Lo T-C, Chen C H, Sawhney A, Agrawal A 2007 Piezoresistive sensors for smart textiles, *Proc. SPIE* **6524** 65241I

[15] Mazzoldi A, de Rossi D, Lorussi F, Scilingo E P, Paradiso R 2002 Smart Textiles for Wearable Motion Captures Systems, *AUTEX Research Journal* **2**, 199-203

[16] Scilingo E P, Lorussi F, Mazzoldi A and de Rossi D 2003 Strain-Sensing Fabrics for Wearable Kinaesthetic-Like Systems, *IEEE Sensors Journal* **3** 460-467

[17] Grillet A, Kinet D, Witt J, Schukar M, Krebber K, Pirotte F and Depré A 2008 Optical Fiber Sensors Embedded Into Medical Textiles for Healthcare Monitoring, *IEEE Sensors Journal* **8** 1215-1222

[18] Harris A J and Castle P F 1986 Bend loss measurements on high numerical aperture single-mode fibers as a function of wavelength and bend radius, *J. Lightwave Technol.* **LT-4** 34-40

[19] Faustini L and Martini G 1997 Bend loss in single mode fibers, *J. Lightwave Techn.* **15** 671-679

[20] Wang Q, Farrell G and Freir T 2005 Theoretical and experimental investigations of macro-bending losses for standard single mode fibers, *Opt. Expr.* **13** 4476-4484

[21] El-Sherif M A 1989 On-fiber sensor and modulator, *IEEE Trans. Instrum. Meas.* **38** 595-598

[22] Tillmanns A, Heimlich F, Brücken A and Weber M O 2009 Weft knitted spacer fabrics as pressure sensors, *Technical Textiles* **52** E207

[23] Meyer J, Arnrich B, Schumm J and Tröster G 2010 Design and Modeling of a Textile Pressure Sensor for Sitting Posture Classification, *IEEE Sensors Journal* **10** 1391-1398

[24] Hertleer C, Grabowska M, van Langenhove L, Catrysse M, Hermans B, Puers R, Kalmar A, van Egmond H and Matthys D 2004 The use of electroconductive textile material for the development of a smart suit, *4th AUTEX Conference Roubaix 2004*

[25] Coosemans J, Hermans B, Puers R 2006 Integrating wireless ECG monitoring in textiles, *Sensors and Actuators A* **130-131** 48-53

[26] Xu P J, Zhang H and Tao X M 2008 Textile-structured electrodes for electrocardiogram, *Textile Progress* **40** 183-213

[27] Silva M, Catarino A, Carvalho H, Rocha A, Monteiro J and Montagna G 2009 Textile sensors for ECG and respiratory frequency on swimsuits, *Intelligent Textiles and Mass customization International Conference Casablanca/Morocco* 301-310

[28] Mühlsteff J, Such O, Schmidt R, Perkuhn M, Reiter H, Lauter J, Thijs J, Musch G and Harris M 2004 Wearable approach for continuous ECG and activity patient-monitoring, *Proc. 26th Ann. Int. IEEE EMBS Conference* 2184-2187

[29] Pacelli M, Loriga G, Taccini N and Paradiso R 2006 Sensing fabrics for monitoring physiological and biomechanical variables: E-textile solutions, *Proc. 3rd IEEE-EMBS Int. Summer School Symp. Med. Dev. Biosens.* 1-4

[30] Habetha J 2006 The MyHeart project – Fighting cardiovascular diseases by prevention and early diagnosis, *Proc. 28th Ann. Int. IEEE EMBS Conference* 6746-6749

[31] Luprano J, Sola J, Dasen S, Koller J M and Chetelat O 2006 Combination f body sensor networks and on-body signal processing algorithms: The practical case of MyHeart project, *Proc. Int. Workshop Wearable Implantable Body Sens. Netw.* 76-79

[32] Luprano J 2006 European projects on smart fabrics, interactive textiles: Sharing opportunities and challenges, *Workshop Wearable Technol. Intel. Textiles, Helsinki/Finland*

[33] Weber J L and Porotte F 2006 Medical remote monitoring with clothes, *Int. Workshop on PHealth, Luzern/Switzerland*

[34] Kim S, Leonhardt S, Zimmermann N, Kranen P, Kensche D, Müller E and Quix C 2008 Influence of contact pressure and moisture on the signal quality of a newly developed textile ECG sensors shirt, *Proc. Of the 5th International Workshop on Wearable and Implantable Body Sensor Networks, Hong Kong/China*

[35] Aumann A, Trummer S, Brücken A, Ehrmann A and Büsgen A 2014 Conceptual design of a sensory shirt for fire-fighters, *Text. Res. J.* **84** 1661-1665

[36] Han J-W, Kim B, Li J and Meyyappan M 2013 A carbon nanotube based ammonia sensor on cotton textile, *Appl. Phys. Lett.* **102** 193104

[37] Liu C, Rousseau I A, Qin H and Mather P T 2002 Tailored Shape Memory Polymers: Not all SMPs are Created Equal, *Proc. of The First World Congress on Biomimetics, Albuquerque/New Mexico*

[38] Lendlein A 2003 Tailor-made intelligent polymers for biomedical applications, in: Tao X (Ed.), *Smart Fibres, Fabrics and Clothing*, Woodhead Publishing Ltd, Cambridge, pp. 278-290

[39] Gall K, Dunn M L, Liu Y, Stefanic G and Balzar D 2004 Internal stress storage in shape memory polymer nanocomposites, *Appl. Phys. Lett.* **85** 290

[40] Winchester R C C and Stylios G K 2003 Designing knitted apparel by engineering the attributes of shape memory alloy, *International Journal of Clothing Science and Technology* **15** 359-366

[41] Tang S L P and Stylios G K 2006 An overview of smart technologies for clothing design and engineering, *International Journal of Clothing Science and Technology* **18** 108-128

[42] Senol Y, Akkan T, Bulgun E Y and Kayacan O 2011 Active T-Shirt, *International Journal of Clothing Science and Technology* **23** 249-257

[43] Berglin L 2005 Spookies: combining smart materials and information technology in an interactive toy, *Proc. of the 2005 conference on Interaction design and children*, 17-23

[44] Hu B, Li D, Ala O, Manandhar P, Fan Q, Kasilingam D and Calvert P D 2010 Textile-Based Flexible Electroluminescent Devices, *Advanced Functional Materials* **21** 305-311

[45] Sloma M, Janczak D, Wroblewski G, Mlozniak A, Jakubowska M 2014 Electroluminescent structures printed on paper and textile elastic substrates, *Circuit World* **40** 13-16

[46] Broianigo F, Haag-Dederichs S, Neumann L, Terzioglu F, Lempa E, Steinem C, Ehrmann A and Rabe M 2011 Das leuchtet ein – Elektrolumineszente Systeme in der textilen Anwendung, *Textilveredlung* 3/4 2011

[47] Hao L, Yi Z, Li C, Li X, Yuxiu W and Yan G 2012 Development and characterization of flexible heating fabric based on conductive filaments, *Measurement* **45** 1855-1865

[48] Kayacan O, Bulgun E and Sahin O 2008 Implementation of Steel-based Fabric Panels in a Heated Garment Design, *Text. Res. J.* **79** 1427-1437

[49] Kayacan O and Bulgun E Y 2009 Heating behaviors of metallic textile structures, *International Journal of Clothing Science and Technology* **21** 127-136

[50] Hamdani S T A, Potluri P and Fernando A 2013 Thermo-Mechanical Behavior of Textile Heating Fabric Based on Silver Coated Polymeric Yarn, *Materials* **6** 1072-1089

[51] Kirstein T, Cottet D, Grzyb J and Tröster G 2002 Textiles for signal transmission in wearables, *Proc. ACM of First Workshop on Electronic Textiles, San Jose/California*
[52] Kuhn H H and Child A D 1998 Electrically conducting textiles, in: Skotheim T A, Elsenbauer R L and Reynolds J R (Eds.), *Handbook of Conducting Polymers,* New York: Marcel Dekker, pp. 993-1013
[53] Lesnikowski J 2011 Textile Transmission Lines in the Modern Textronic Clothes, *Fibres & Textiles in Eastern Europe* **19**, 89-93
[54] Locher I, Klemm M, Kirstein T and Tröster G 2006 Design and characterization of purely textile patch antennas, *IEEE Transactions on Advanced Packaging* **29** 777-788
[55] Jost K, Dion K and Gogotsi Y 2014 Textile energy storage in perspective, *Journal of Materials Chemistry A* **2** 10776-10787
[56] Jost K, Perez C R, McDonough J K, Presser V, Heon M, Dion G and Gogotsi Y 2011 Carbon coated textiles for flexible energy storage, *Energy Environ. Sci.* **4** 5060
[57] Gorgutsa S, Gu J F and Skorobogatiy M 2012 A woven 2D touchpad sensor and a 1D slide sensor using soft capacitor fibers, *Smart Mater. Struct.* **21** 015010
[58] Jost K, Stenger D, Perez C R, McDonough J K, Lian K, Gogotsi Y and Dion G 2013 Knitted and screen printed carbon fiber textile-supercapacitors for applications in wearable electronics, *Energy Environ. Sci.* **6** 2698-2705
[59] Liu Y, Gorgutsa S, Santato C and Skorobogatiy M 2012 Flexible, Solid Electrolyte-Based Lithium Battery Composed of $LiFePO_4$ Cathode and $Li_4Ti_5O_{12}$ Anode for Applications in Smart Textiles, *J. Electrochem. Soc.* **159** A349-A356
[60] Jost K, Durkin D P, Haverhals L M, Brown E K, Langenstein M, de Long H C, Trulove P C, Gogotsi Y and Dion G 2014 Natural Fiber Welded Electrode Yarns for Knittable Textile Supercapacitors, *Adv. Energy Mater.* **2014** 1401286

IOP Conf. Series: Materials Science and Engineering **141** (2016) 012009 doi:10.1088/1757-899X/141/1/012009

Thermal comfort of diving dry suit with the use of the warp-knitted fabric

I Lenfeldova[1], **L Hes**[2] **and M Annayeva**[2]
[1] Department of Textiles Technologies, Faculty of Textile Engineering, Technical University of Liberec, Studenska 2, 46117 Czech Republic
[2] Department of Textile Evaluation, Faculty of Textile Engineering, Technical University of Liberec, Studenska 2, 46117 Czech Republic

Abstract. Achievement of a good level of thermal comfort of under-suits for dry suit diving which enable also the required mobility of the diver in water is inevitable not only for the scuba sport and commercial diving people but also for safety and activities of people who make research under water. The aim of this work is to verify whether selected knitted structures (which are not waterproof) can substitute the currently used textile materials (nonwovens). This dry-suit innovation is intended to increase the properties which correspond to the perception of thermal comfort of the diver in water. To achieve this objective, the Alambeta thermal tester was used in the study for experimental determination of thermal resistance of spacer warp knitted fabric at varying contact pressure. The studied textiles were expected to be very suitable for the intended application due to their low compressibility which yields relatively high thickness a hence increased thermal insulation.

1. Introduction

A Spacer fabric is a double-faced fabric knitted on a double needle bar machine. The distance between the two surfaces is retained after compression by the resilience of the pile yarn (usually mono-filament) that passes between them. End-used for spacer fabrics include moulded bra cups, padding, and linings. Other applications are being investigated [1].

To analyse the effect of fabric thermal properties under high value of pressure, it should be noted that a spacer knitted fabric has a different mechanical behaviour in different directions. The various constructions between the wale-wise and course-wise directions can cause different fabric compression behaviour due to the various number of contacting points. In addition, during compression the spacer monofilaments into the structure can contact one another [2].

It is necessary to point out that the compression resistance of spacer fabrics increases with decreasing the inclination of the spacer yarns, because the spacer yarns with lower inclination are more oriented to the direction of the impact. This can be easily understood by referring to the theory of elastic stability from which the critical load P_{cr} of an elastic rod is given by the following equations:

when the two ends of the elastic rod are pinned: Firstly,

$$P_{cr} = \frac{\pi^2 EI}{l^2} \quad (1)$$

And secondly, when the two ends are fixed:

$$P_{cr} = \frac{4\pi^2 EI}{l^2} \quad (2)$$

Where E is the module of elasticity of the material, I is the moment of inertia of the cross-sectional area of the rod, and l is the length of the elastic rod. From Equations (1) and (2), it is easily shown that the shorter elastic rod exhibits the higher critical load, which leads to higher compression resistance. Therefore, the fabric with lower angle of inclination will have higher compression resistance [3].
With increasing the fabric thickness, the spacer yarns get longer. If considering the spacer yarns as slender rods, their compression resistance will decrease with increasing their length, as indicated by Equations (1) and (2).

2. Thermal properties of textiles

Thermal comfort properties of textile materials have gained the attention of researchers in recent times. Although a plethora of researches have been conducted on the mechanical properties of textile fabrics, they have hardly played any role during the actual use of the fabrics. In contrast, comfort properties determine the way in which the heat, air and water vapour are transmitted across the fabric. During heavy activities, the body produces lots of heat energy and the body temperature rises. To reduce the temperature, the body perspires in liquid and vapour form. When this perspiration is evaporated to atmosphere, the body temperature reduces.

2.1Principle of ALAMBETA instrument – fabric thermal properties tester

This apparatus used in this study enables the measurement of the following thermal parameters: thermal conductivity, thermal absorbtivity, thermal resistance and sample thickness. The Alambeta simulates the dry human skin and its principle depends in mathematical processing of time course of heat flow passing through the tested fabric due to different temperatures of bottom measuring plate (22°C) and measuring head (32°C). When the specimen is inserted, the measuring head drops down, touches the fabrics and the heat flow levels are processed in the computer and thermo-physical properties of the measured specimen are evaluated [4].
Thermal properties of textiles such as the thermal resistance, thermal conductivity and thermal absorbtivity are influenced by the yarn composition and structure, fabric structure, density, humidity, type of textile construction, surface treatment, filling and compressibility, air permeability, surrounding temperature and other factors. Thermal conductivity coefficient λ presents the amount of heat, which passes from 1m^2 area of material through the distance 1 m within 1 s and creates the temperature difference 1 K. The highest thermal conductivity exhibit metals, whereas polymers have low thermal conductivity, ranging from 0,2 to 0,4 W/m/K.
Thermal conductivity of textile structures generally reaches levels from 0,033 to 0,01 W/m/K. Thermal absorbtivity b of fabrics was introduced by Hes [4] to characterise thermal feeling (heat flow level) during short contact of human skin with the fabric surface. Providing that the time of heat contact τ between the human skin and the textile is shorter than several seconds, the measured fabric can be simplified into semi-infinite homogenous mass with certain thermal capacity ρc (J/m^3) and initial temperature t^2. The higher is thermal absorbtivity of the fabric; the cooler is its feeling. In the textile praxis, this parameter ranges from 20 Ws$^{1/2}$/m^2 /K for fine nonwoven webs to 600 Ws$^{1/2}$/m^2 /K for heavy wet fabrics. Thermal resistance R (m^{2}K/W) depends on fabric thickness h and thermal conductivity λ, i.e. the equation is given by relation:

$$R = h/\lambda \tag{3}$$

3. Experimental results

Three types of the dive dry-suits (NW – 1, 2, 3, 3 – used) were used for non-destructive method of ALAMBETA testing. The basic characteristic parameters of these nonwoven structures were not found. It was not possible to cut the real suit for studying.
For the comparison, thermal parameters of the non-used dive-dry suit and dry-suit after 300 scuba dives were evaluated (NW – 3, 3 – used). Afterwards, 6 different warp-knitted spacer fabrics and 4 types of nonwoven fabrics were experimentally evaluated under laboratory conditions (t = 23 °C and φ = 48 %).

3.1Testing of the real dive dry-suits

The first results of the ALAMBETA instrument are given in the Table 1 add in the Figures 1 – 3.

Table 1. Thermal parameters of the evaluated dive-dry suits

	h (mm)	10^3 λ (W/mK)	10^3 R (m².K/W)	q (W/m²)	b (Wm^{-2}.s$^{1/2}$.K^{-1})
NW - 1	6.64	41.6	159	0.119	57.8
NW - 2	12.4	49.4	250	0.178	55
NW - 3	11.5	41.8	275	0.126	39.4
NW - 3 (used)	8.65	40.6	213	0.158	48.5

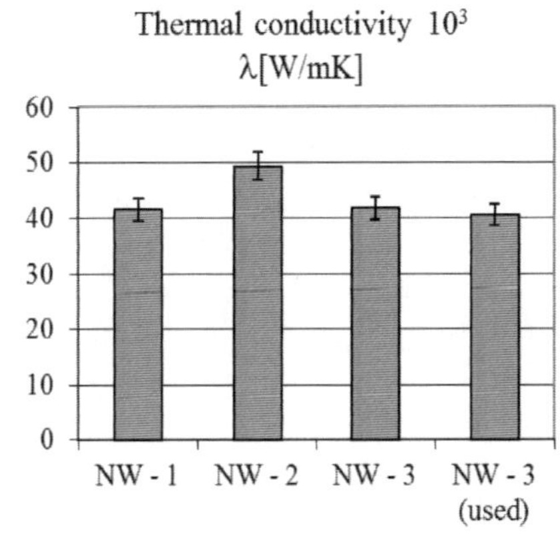

Figure 1. Thermal conductivity of the dive dry-suit

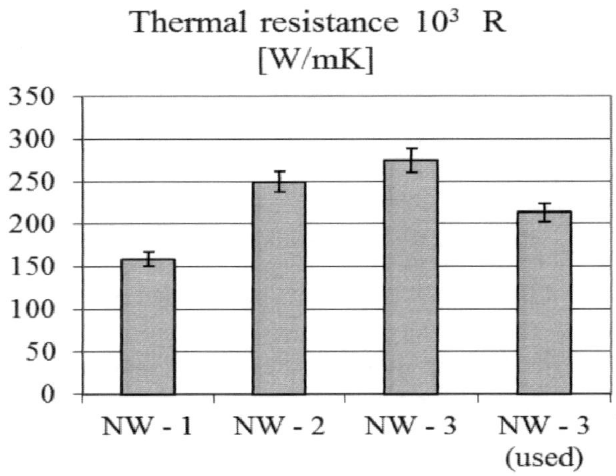

Figure 2. Thermal resistance of the dive dry-suit

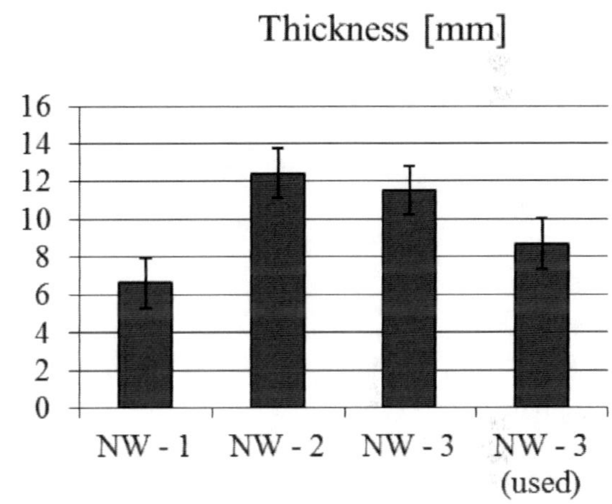

Figure 3. Thickness of the dive dry-suit (ALAMBETA)

Thermal conductivity values of all dry-suits exhibited low variability, in respect of their confidence interval (Figure 1). The personal experience of the scuba divers refers to the use of the suit named NW – 3. They consider that dry-suit as more comfortable. The determined thermal resistance levels correspond to the thickness of the suits (Figures 2, 3). The structure of these dry-suits was not analysed in order to avoid their destruction.

Repeated diving cycles bring the decrease of thermal resistance of the suits, as it depends on the fabric thickness. The biggest drop of the thickness of the used dry-suits is the main reason why this suit is not comfortable for diving in cold water. The water pressure during the diving brings about the changes of the nonwoven structure.

3.2 Testing of the cut samples of innovative dry-suit by mans of the ALAMBETA instrument

In the experiment five structures of dry-suit materials were compared (Table 2), four based on conventional nonwovens (now used and produced) and one with the innovative warp-knitted spacer layer (Figure 4). Samples consist from:

- Outer layer – woven fabric
- Middle layer – nonwoven fabric or warp-knitted spacer fabric
- Inner layer – warp knit or weft knit (in the case of warp-knitted spacer the inner layer was not necessary)

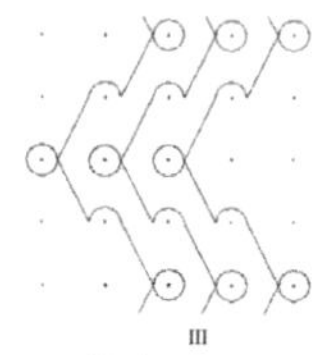

Figure. 4 Notation of the spacer monofilament (GB 3 and 4), PL 76 tex f 24, PL 33 f1.

Table 2. Thermal comfort properties of all studied dry-suits. (The values of individual suits were determined under the pressure 200 and 1000 Pascal).

Type of used material for simulation of the dry-suits	$10^3\ \lambda$ (W/mK)	$10^3\ R$ (m².K/W)	q (W/m.m)	b (Wm^{-2}. s$^{1/2}$.K^{-1})	h (mm)
1=Woven/Nonwoven (hollow fibre) 150 g/m²/Warp knit	50.1	253	0.086	17	12.7
1- 1000 Pa	38.2	124	0.127	30.1	4.74
2=Woven/Nonwoven (hollow fibre +Ag) 150 g/m²/Warp	47.9	327	0.053	15	15.7
2- 1000 Pa	38.5	192	0.076	18.7	7.41
3=Woven/Nonwoven (hollow fibre) 150 g/m²/Weft knit	46.7	227	0.093	19	10.6
3- 1000 Pa	37.4	124	0.114	31.9	4.63
4=Woven/Nonwoven (profile fibre) 300 g/m²/Warp knit	49.3	404	0.024	5.9	19.9
4- 1000 Pa	40.4	215	0.033	11.8	8.69
5=Woven/Warp-knited spacer	49.9	76.2	0.302	58.8	3.8
5- 1000 Pa	47.7	73.2	0.328	63.9	3.49

The measurement was performed under the two pressure levels which are possible to set in the ALAMBETA tester. The confidence interval of thermal conductivity and thickness (10 samples) was also determined; see the Figure 5, 6.

Unfortunately, better thermal properties did not result from the use of multiple layers of nonwoven fabrics (see the results of samples No. 4). The disadvantage of the increasing the mass of the dry-suits will be higher than the advantage of the comfort benefit.

As expected, under the increasing pressure the thickness of the spacer fabric decreases – see the sample 5. However, thermal resistance of the warp spacer knits suits less decreases under high pressure then thermal resistance of the suits made of nonwoven fabrics including fabrics with special hollow fibres.

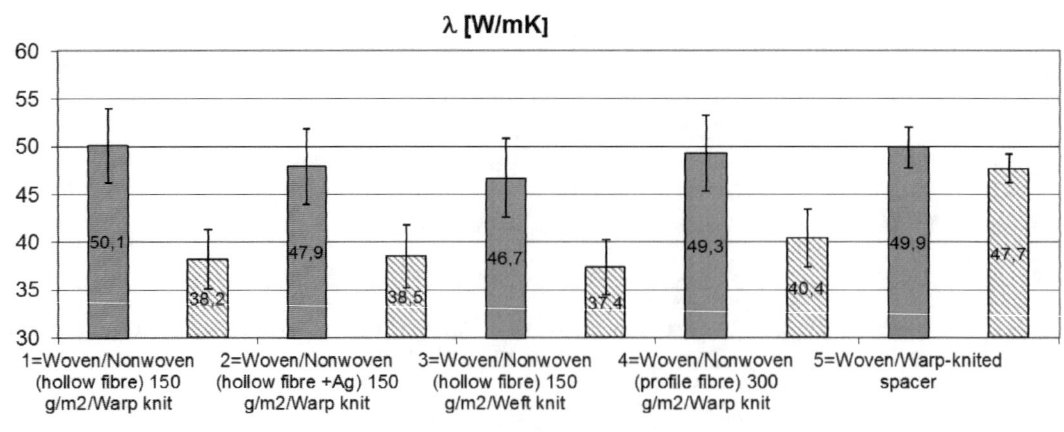

Figure 5. Thermal conductivity of the cut samples – two levels of pressure

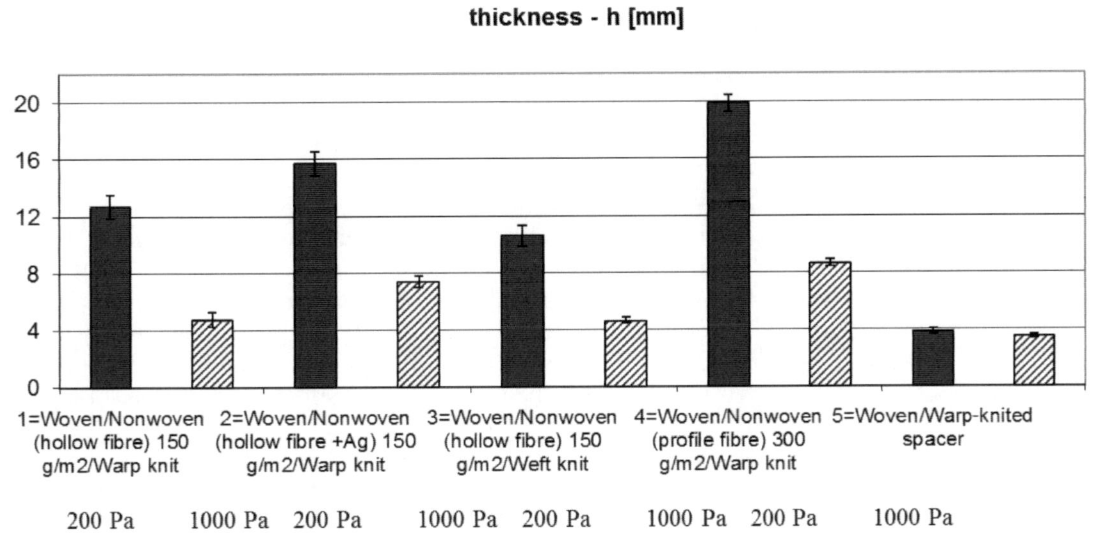

Figure 6. Thickness of the cut samples (ALAMBETA) – two levels of pressure

3.3 Measurement of the inclination spacer fibres of warp-knits – CT scanner

Warp-knitted spacer fabrics are often used for improvement of the compression behaviour which is depends on the three layer structure configuration (lapping), on the spacer materials (monofilament diameter) and on the stitch density of two layers (outer). Thanks to their small inclination at the fabric compression these materials enable certain air flow inside the fabric and cause certain thermal comfort, which depends on the thickness during the contact pressure that is on the compression resistance.

To determine the real value of the inclination of the spacer fibres inside the warp-knitted fabric and the impact of the fixation process on the change of the monofilament inclination (the samples thickness has to be changed too) the experiment with CT scanner was performed.

CT scanner uses an x-ray source with adjustable voltage and a range of filters for versatile adaptation to different object densities. Internal structures are reconstructed as a series of 2D cross sections (Figure 7) which are then used to analyse the two and three dimensional morphological parameters of the object. The process is non-destructive [5].

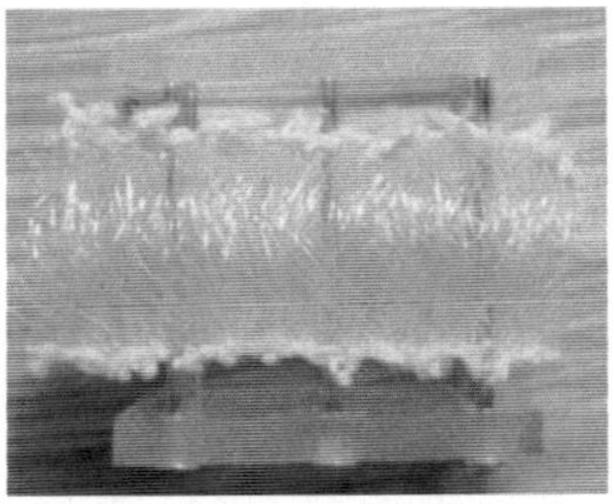

Figure 7. Apparatus for micro CT-scanner SkyScan 1174

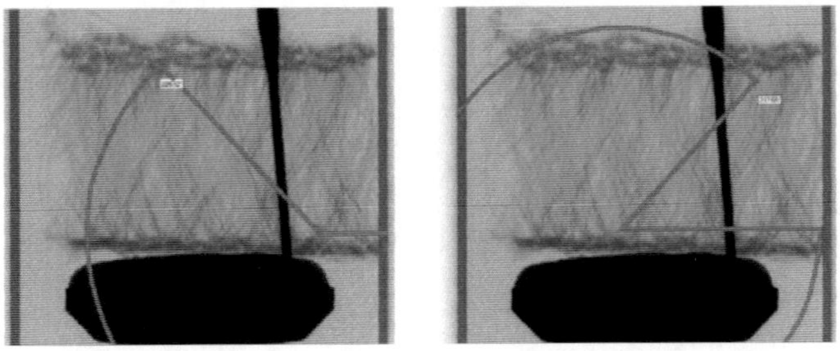

Figure 8. Measurement of the spacer monofil angles

This process starts by the approximation of one of the monofilaments and then by approximation of another monofilament, which crosses the first one. Two angles are then determined and after subtraction the real inclination of two monofilament axes are calculated. Statistical data of the angles from the three samples (thirty values) are given in the Table 3.

Table 3. Inclinations of the monofilaments axes from the spacer warp-knitted fabric surface [6].

	Before fixation	After fixation
α (°)	58.8	91.8
s (°)	8.16	66.66
s^2 (°)	66.6	44.3
v (%)	13.8	7.25
Confidence (°)	55.7-61.8	89.3-94.3

If the angle of two monofilaments is high (after fixation), the resistance to compression of these structure is lower. The process of fixation changes the structural parameters and these changes increase the pressure resistance. Higher monofilament angle predicts a structure with higher compressibility. The other characteristics which affect the compressibility of spacer fabrics are density of spacer warp-knitted fabric, type and count of spacer fibres and lapping diagram together with the thickness of the fabric, which is given by the needle bed distance of the double needle bar raschel machine.

The structure of the double needle bar warp-knitted filet in its cross section is given by the lapping diagram of the monofilaments. During the fabric production and before fixation process the monofilament angle is in the non-relaxed state and the fabric mechanical characteristics (which depend on the thickness and compressibility) can differ in values.

3.4 Innovation of the dive dry-suit with the spacer knits

As follows from the Tab. 4, thermal resistance of the warp spacer knits suits under high pressures practically did not change, contrary to the thermal resistance of the suits made of nonwoven fabrics including fabrics with special hollow fibres, which dropped substantively.

This resistance against pressure of the suit based on spacer fabric is caused by high compression resistance of the middle layer. The inclination of the spacer monofilament fibres brings this stability.

Some characteristics and results of the thickness testing especially of the spacer knits are given in the Table 4. It is probably the first attempt to describe the thermal comfort of the spacer fabrics. The samples differ in their thickness (and therefore the materials for spacer layer differ too). The notation of these structures varies and the threading of the guide bar too.

During the samples compression the yarn geometry changes and fabric structure is temporarily or permanently disrupted. Owing to this process the thermal resistance decreases. The compression level Z can be determined by the relation and may also be formatted as

$$Z = \frac{h_1 - h_2}{\log p_2 - \log p_1} \tag{4}$$

where h_1 (mm) is thickness under the pressure p_1 (Pa) and h_2 (mm) is thickness under the pressure p_2 (Pa). These values were determined with UNI-THICK instrument.

Table 4. Characteristics of the tested materials

	Material	Thickness	Square mass	Thickness difference Δh (%) under pressure		
		(mm)	(gm^{-2})	(200 - 1000) (Pa)	(1000 - 10000) (Pa)	(10000 - 100000) (Pa)
Spacer A	PL 83 tex, f180; PL 0.25mm	10	500	5.37	45.44	80.22
Spacer B	PL 7.6 tex, f 24; PL 0.09 mm	6	300	4.5	33.45	81.47
Spacer C	PL 83 tex, f180; PL 0.25mm	6	335	5.3	39.1	77.8
Spacer D	PL 83 tex, f180; PL 0.25mm	7	490	3.94	23.7	69.93
Spacer E	PL 83 tex, f180; PL 0.25mm	10	850	6.78	20.59	65.93
Spacer F	PL 83 tex, f180; PL 0.25mm	10	800	5.64	16.37	63.18
Nonwoven	PL	18.87	54.09	18.87	54.09	87.13

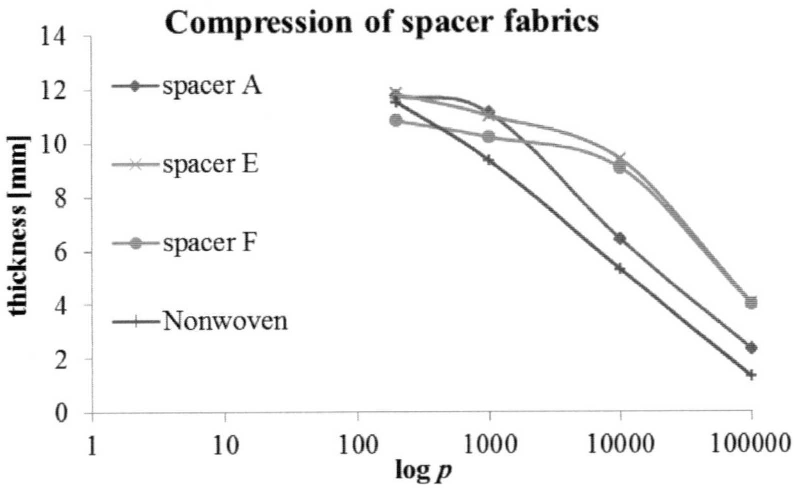

Figure 9. The thickness influence of warp-knitted spacer fabrics and nonwoven

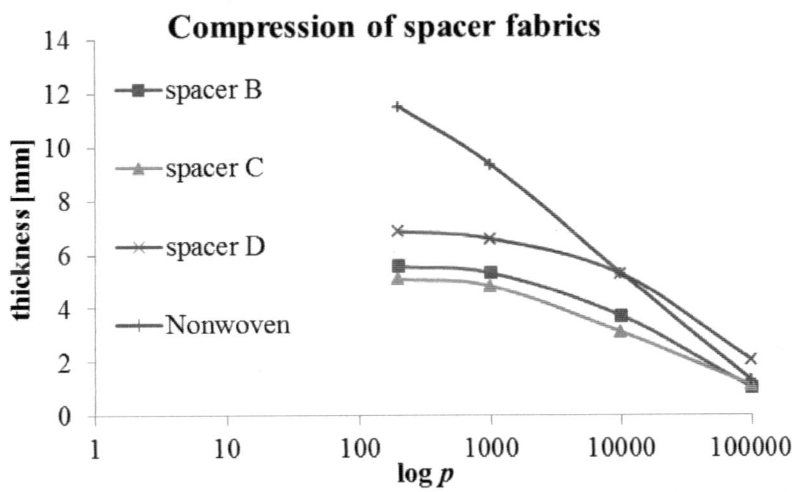

Figure 10. The effect of pressure on the thickness of warp-knitted spacer fabrics and nonwovens

The Figure 10 clearly illustrates the thickness changes. Thermal conductivity, thermal resistance and thermal absorptivity depend on the changed volume density (porosity) of the samples.

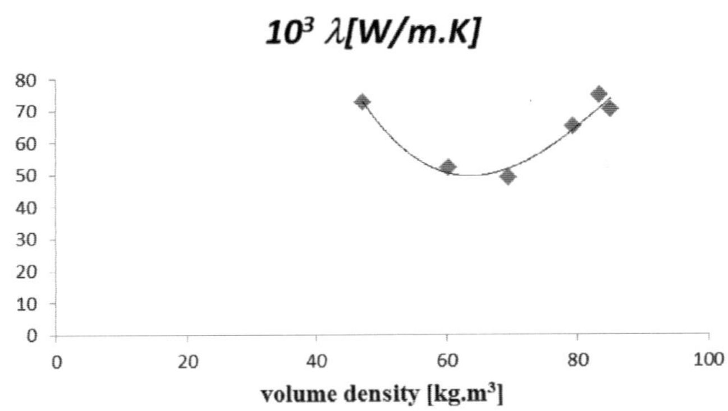

Figure 11. Thermal conductivity of spacer knits and nonwovens

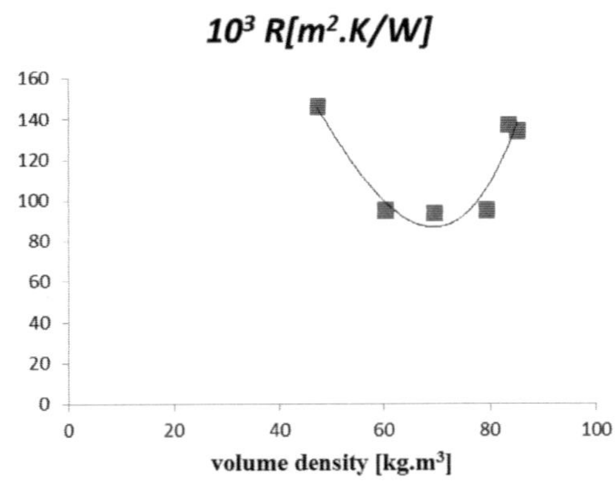

Figure 12. Thermal resistance of the spacer knits and nonwovens

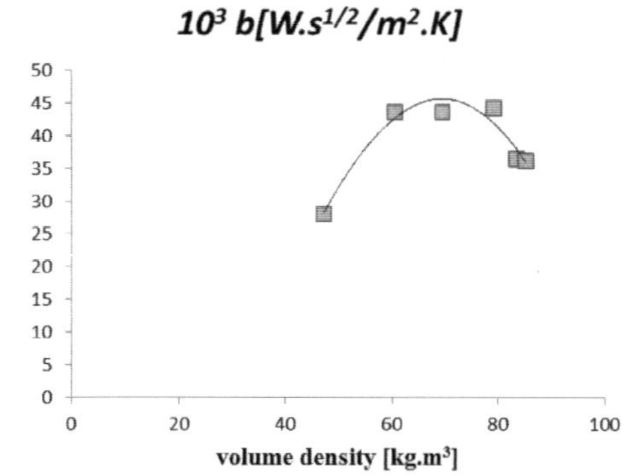

Figure 13. Thermal absorptivity of spacer knits and nonwovens

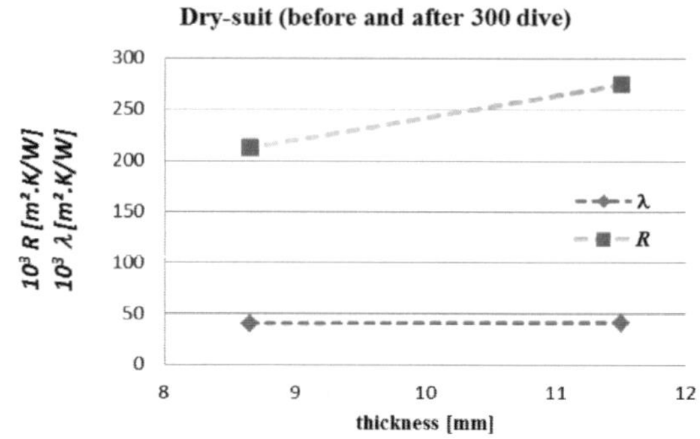

Figure 14. Comparison of properties a new and used dive dry-suit with nonwovens (influence of water pressure)

4. Results and discussion

The increasing water pressure during the scuba diving influences the structure of dry-suits and changes their thermal properties, as can be seen in the Figure 14.

From the above results follow, that this sort of suits can really offer the expected high level of thermal insulation by maintaining an air (or special gas) layer between the body and the cold water. On the other hand, it features certain disadvantage resulting from higher bending and shearing stiffness, which may reduce the mobility of the diver under water. This disadvantage can be not so inconvenient due to the position in the water – trim.

The advantages of the innovative dry-suit with spacer warp knitted fabric are:

- better thermal properties which are necessary for scuba-diver comfort under cold water
- due to the pressure resistant the thickness of the suit will not be changed after diver cycles.

Contrary to these benefits the disadvantages with using spacer knits are:

- smaller un-comfortableness in "swimming" (stiffness of the knitted textiles)
- problems with diving down owing to huge volume of the air into the structure
- difficulty in processing of spacer fabric into a dive-dry suit.

5. References

[1] Spencer D 2001 *Knitting technology a comprehensive handbook and practical guide* (Woodhead Publishing Limited) p 357-77

[2] Liu Y and Hu 2014 *An Experimental Study of Compression Behavior of Warp-knitted Spacer Fabric* (North Carolina State University) J. Eng. Fibers Fabr. **vol 9** pp 61-9

[3] Liu Y and Hu 2012 *Impact compressive behaviour of warp-knitted spacer fabrics for protective applications* Textile Res. J. **vol 12 pp** 774 – 88

[4] Hes L, Araujo M and Djulay V 1996 *Effect of Mutual Bonding of Textile Layers on Thermal Insulation and Thermal-Contact Properties of Fabric Assemblies* (Text Res J 66) pp 245-50

[5] *SkyScan 1174 Micro-CT* [online]. [cit. 2016-02-15]. Available from www : <https://www.microphotonics.com/products/micro-ct/skyscan-1174-micro-ct/>,

[6] Vancura F 2010 *Influence angle laying monofilaments on resistance in the fface of compression double bar warp knitted* fabrics – *Spacer fabrics* (Liberec: Technical University of Liberec Thesis in Czech)

[7] Jablonski J 2006 *Doing it Right: The Fundamentals of Better Diving* (Global Underwater Explorers)

[8] *Diving weighting system* [online]. 14.1.2016 [cit. 2016-02-13]. Available from www : < https://en.wikipedia.org/wiki/Diving_weighting_system>

IOP Conf. Series: Materials Science and Engineering **141** (2016) 012010 doi:10.1088/1757-899X/141/1/012010

About the automated pattern creation of 3D jacquard double needle bed warp knitted structures

W Renkens[1], Y Kyosev[2]

[1]Renkens Consulting, Aachen, Germany

[2]Hochschule Niederrhein – University of Applied Sciences, Research Institute for Textile and Clothing, 41065 Mönchengladbach, Germany

W.Renkens@gmx.de

Abstract. Three dimensional structures can be produced on jacquard warp knitting machines with double needle bed. This work presents theoretical considerations about the modelling and simulation of these structures. After that a method is described, how to obtain production parameters from the simulation data. The analysis demonstrates, that the automated pattern creation of 3D structures is not always possible and not all mathematical solutions of the problem can be knittable.

1. Introduction
The jacquard warp knitting machines were initially used for production of samples with different decorative effects. In the last decade their control and programming allow as well the production of almost seamless clothing and other warp knitted pieces with tubular or another three dimensional form [1, 2]. The development and the production of such fabrics requires very good understanding of the warp knitting process and pattern and suitable CAD software for designing and manipulating the pattern before the production. This work does not concentrate on the modelling of the 3D configuration of the yarns axis in the structure, it considers the structure at one lever higher – at the macro level. The main question at that level is how one complex 3D shape can be split into several regions and after that programmed on the warp knitting machine, which in generally produce flat structural elements on its both needle beds and is able to make connection between these beds.

2. State of the art
In the area of the modelling of the yarn geometry of the warp knitted structures are existing several investigations and models [3–9], but the problem of the decomposition of one 3D profile into several parts and the generation of knitted program is very less investigated. In [10] are presented general consideration about the modelling of the loops in jacquard warp knitted fabrics. The modelling of jacquard towels is presented in [11]. The principles of the separation of the areas of different interlacement of the yarns for spacer warp knitted fabrics and the formation of the matrices for the guide control in the case of shoes is given in [12].

3. Patterning for Jacquard Warp Knitting

A simple jacquard warp knitted structure can be based on vertical courses (Figure 2, left), built of lock-stitch (0-1//) and weft yarns (0-0/1-1//), where the weft yarns in some places make longer (0-0/2-2//) or no (1-1/1-1//) underlaps up to the next course.

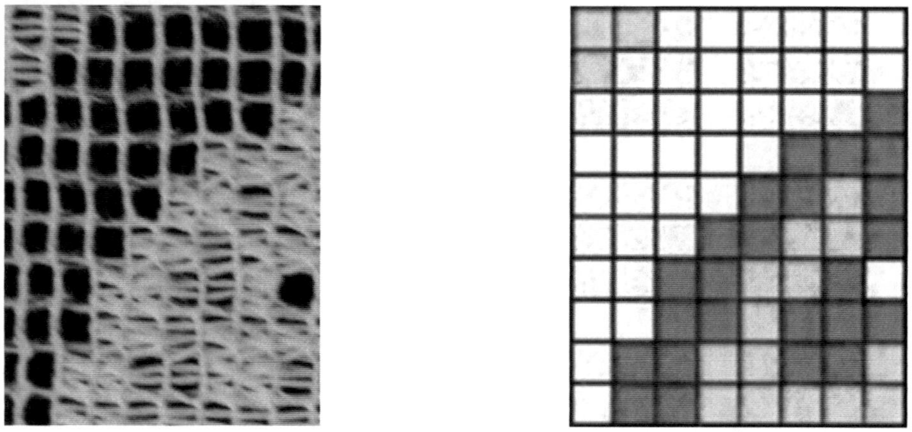

a b

Figure 1. Segment of a fabric (a) and the corresponding colour pattern (b)

There are three possible appearances of the fabric, depending on the number of the yarns in a unit cell formed by two courses and the jacquard needles, transparent areas (no yarns, white), not so dense areas (two yarns, green) and dense areas (four yarns, red). These unit cells are transferred to a colour pattern (figure 1, right). Jiang, Zhang, et.al reported about an automated system for image recognition and classification of the areas, based on wavelet transformation [9]. These colour pattern can be converted quite easily to control data for the machine.

The possibilities on 2 needle-bed machines are more complex. To start with a relative simple case, the right edge of a hollow structure can be considered. In this example, you can distinguish 3 different elements: 1 - front, 2 - left edge and 3 - back (Figure 2a). In order a colour pattern to be created, the model has to be cut at the left front corner and flip the backside to the right, which is common technique for the construction of clothing around human body.

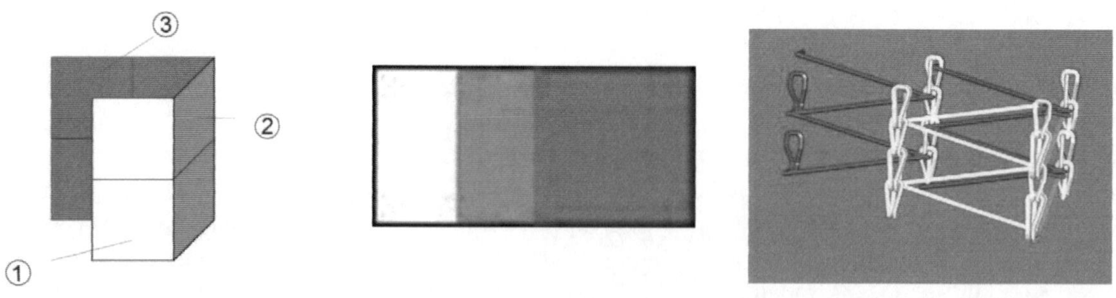

a b c

Figure 2. Model of a left edge (a), corresponding colour pattern (b) and the simulated fabric (c)

The colour pattern can be regarded as a matrix MC(p, g). The rows represent the courses (p) and the columns the guides (g).

IOP Conf. Series: Materials Science and Engineering **141** (2016) 012010 doi:10.1088/1757-899X/141/1/012010

$$MC(p,g) = \begin{bmatrix} c_{1,1} & \cdots & c_{1,g} \\ \vdots & \ddots & \vdots \\ c_{p,1} & \cdots & c_{p,g} \end{bmatrix}$$

The algorithm to compute the simulation requires a Matrix MM(q, g), with q = 8p.

$$MM(q,g) = \begin{bmatrix} m_{1,1} & \cdots & m_{1,g} \\ \vdots & \ddots & \vdots \\ m_{q,1} & \cdots & m_{q,g} \end{bmatrix}$$

Again each column represents a Jacquard-guide and the rows contain the underlap and overlap movement for one unit cell. This movement *m* (8 values) is taken from a look-up-table.

4. Simulating for two needle bed jacquard warp knitting

Starting from sketch of the fabric, several steps have to be performed in order the machine control data to be generated in automated way:

1. Create 2D sketch of the fabric to be produced.
2. Generate a 3D model with existing elements.
3. Generate a matrix with the element or for more intuitive use - image in .BMP-format.
   The front is on the left side and the rear side is mirrored on the right.
4. For each thread create a list of movements for the guides (yarns) through the needles (chain-link).
5. Compute the 3D-Geometry of the warp knitted structure.

Steps 1 to 3 have to be done by the designer manually or using some standard CAD system with 3D capabilities. Step 4 was implemented within MATLAB® and presented in [10]. The principle, implementation and different aspects of performing of Step 5 are reported in several papers, as for instance [1, 8].

In figure 3 these steps are illustrated with a simplified example of an alien glove with a thumb and two 'fingers'. Figure 4 presents the simulated warp knitted structure for this example.

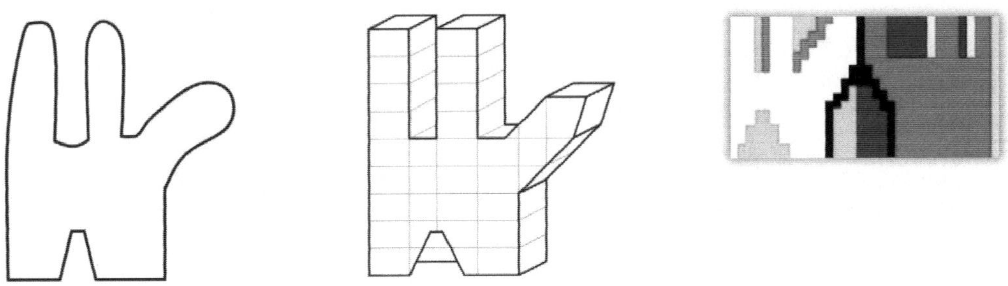

a b c

Figure 3. Alien Glove: Sketch (a), 3D model (b), colour pattern (c).

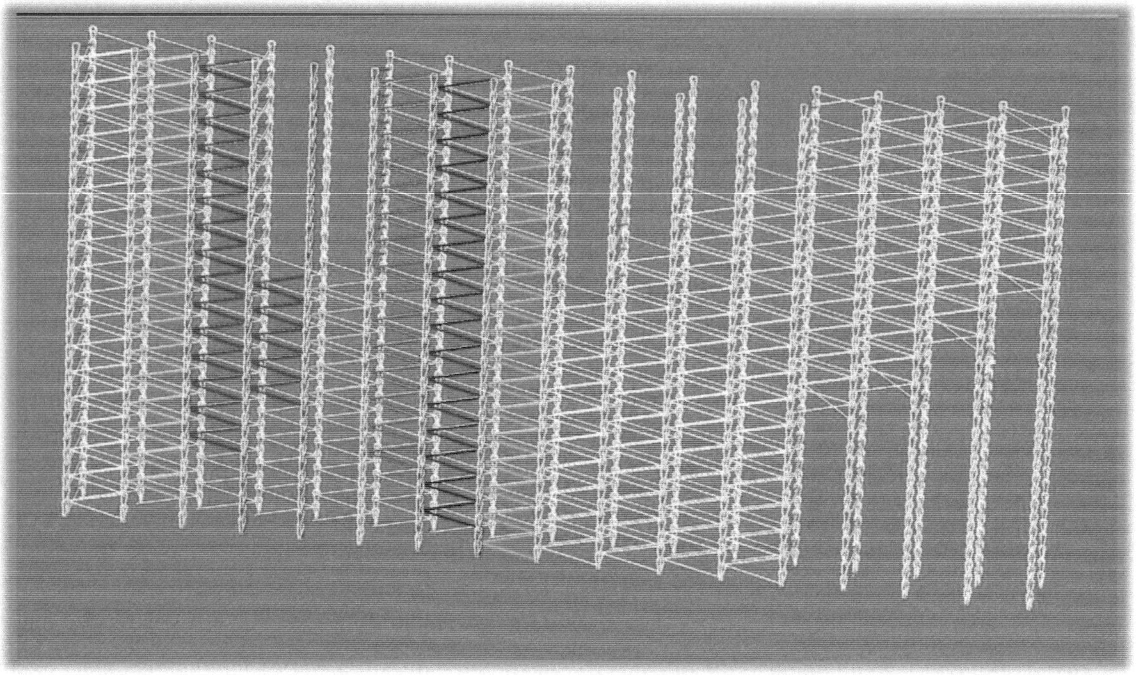

Figure 4. Alien Glove: Simulation

5. Transferring to a jacquard warp knitting machine

Jacquard warp knitting machines have one or more jacquard guide bars, whose guide needles can stay at its position or can move to one position to the right if a command is given (Figure 5 a). Earlier realizations used classical mechanics [13] and the modern solution uses the Piezo-effect [14]. Such individual motions of the single guides allows creation of complex samples, which are used on curtain machines to create pictures on the structures (Figure 1a), on lace machines to produce ground variations, on two needle bed machines to produce 'ready to wear' bodies or hollow structures like gloves or tubes.

The movement of the threads has to be given in absolute values to compute the simulation. These absolute values contain the position of the guides on the Jacquard bar. The control unit of the Jacquard-machine demands two different set of control-data; one for the shogging movement of the jacquard bar (in former times known as pattern disc) and another set of data to control the individual movement of each jacquard guide (in former times given on punched cards). In order to automate the design process, some algorithms have to be developed which take the guide motion as input and generate data for both shogging movement and high-low information for the Jacquard guides.

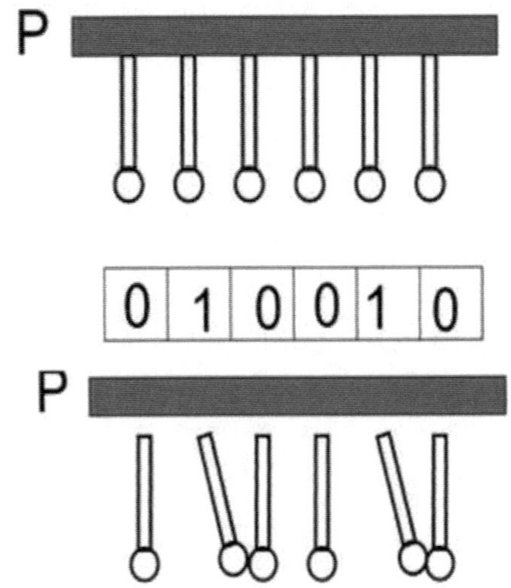

a

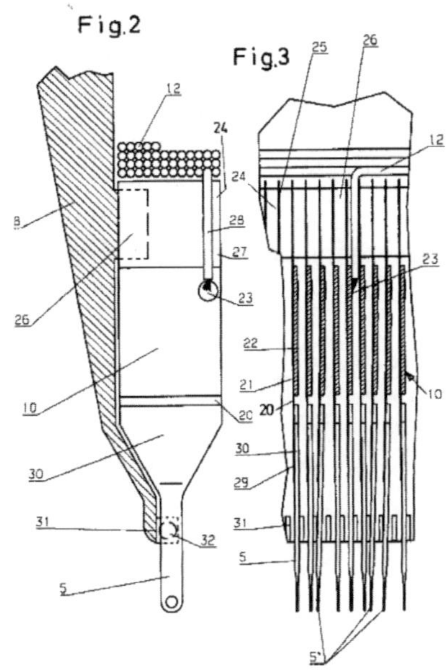

b

Figure 5. Principle of Jacquard movement (a) and the realization on modern jacquard warp knitting machine (b; Patent for the piezo guide bars [14])

6. Problems

Not every geometry can be transferred to warp knitting program for the currently available machines. Figures 6 and 7 visualize one problem during the creation of the warp knitting program. If at a given place two guides have to be "missing", in order for instance to produce a hole at that place, then this cannot be done with bars, placed at one guide bar, because these can move only to one direction (Figure 6b). If in the future the piezo material can be placed on the both sides of the metal guide probably such guide can move to left and right and could be able to solve this problem, but at the current technique this is not possible.

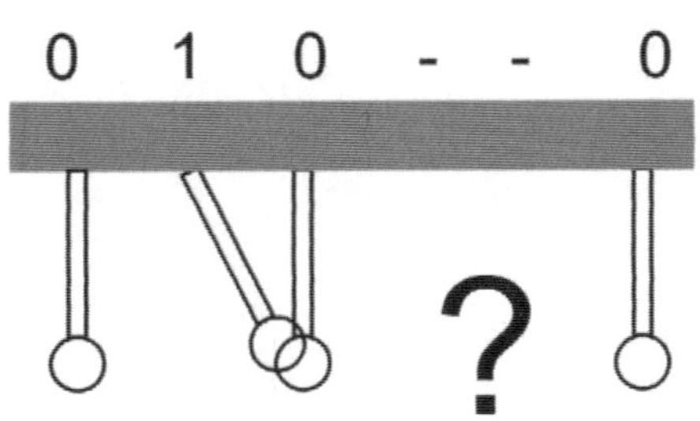

a

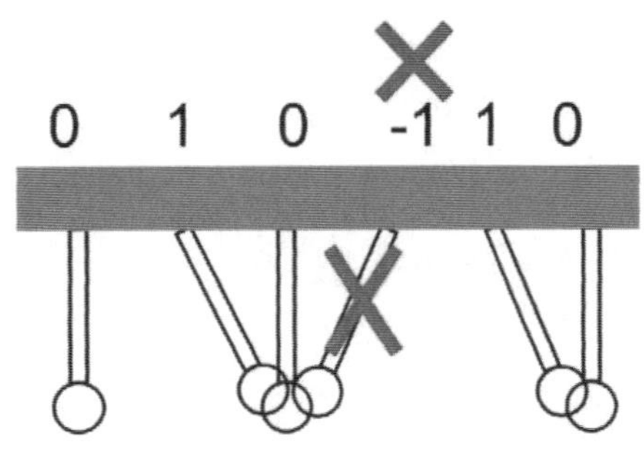

b

Figure 6. How to produce a hole in the area of two guides, if these are required during the remaining time of the process (a)? If motion in the opposite direction were possible (b), then moving one guide to left and one to right would be a solution, but the current piezo guide bars can move only at one direction.

Possible solution of this problem will be to distribute the problematic guides on a separated guide bars (Figure 7b). In this case the single guide can be moved with the complete bar at a different position and back on the same place at later cycles. This solution has its disadvantage – for more complex structures it can happens, that a large set of guide bars with single guides are required, which can become impossible to be produced from the technological point of view. In such case the structure has to be simplified and adjusted, so that it becomes knittable. But this is a demonstration, why the fully automated production of a warp knitting program from 3D data from the profile cannot always have technologically suitable solution.

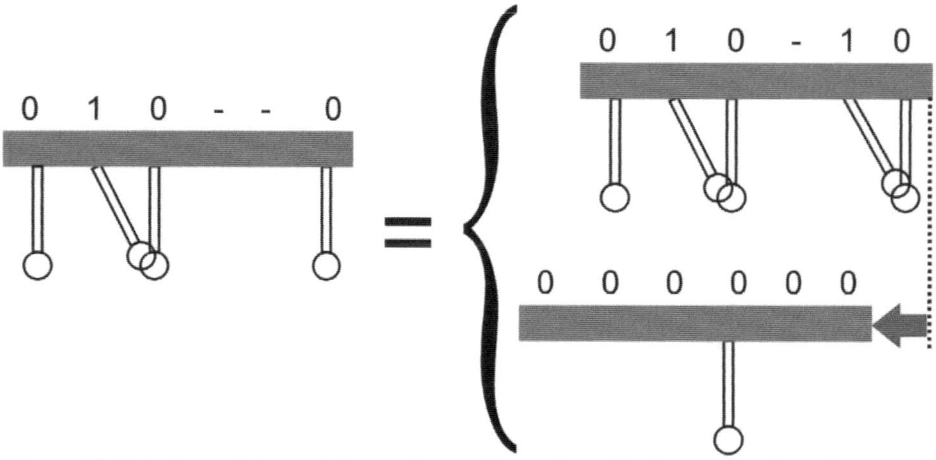

Figure 7. Solution of the problem – the problematic guide has to be placed on another guide bar which can be moved as well.

7. Conclusions

This overview presents a possible method for simulation of 3D Jacquard warp knitted structures and generation of the machine data for production of these. The simulation uses fast and simple geometrical models which can be used only to generate initial geometry of the structures for mechanical simulations, but which is enough to give a rough idea about the fabric during the design process. The explained algorithm make clear which tasks has to be solved in the future in order automated design process from sketch to machine data to be possible.

References

[1] Renkens W, Kyosev Y 2011 Geometry modelling of warp knitted fabrics with 3D form, *Textile Research Journal* **81**(4), 437-443. doi: 10.1177/0040517510385171

[2] Renkens W, Kyosev Y 2007 Geometrical modelling of warp knitted fabrics. In: "Finite Element Modelling of Textiles and Textile Composites", St-Petersburg, 26-28 Sept. 2007, CD-ROM Proceedings

[3] Hart K, Jong S de, Postle R 1985 Analysis of the Single Bar Warp Knitted Structure Using an Energy Minimization Technique: Part I: Theoretical Development, *Textile Research Journal* **55**(8), 489-498. doi: 10.1177/004051758505500807

[4] Hart K, Jong S de, Postle R 1985 Analysis of the Single Bar Warp Knitted Structure Using an Energy Minimization Technique: Part II: Results and Comparison with Woven and Weft Knitted Analysis. *Textile Research Journal* **55**(9), 530-539. doi: 10.1177/004051758505500903

[5] Goktepe O, Harlock S 2002 Three-Dimensional Computer Modeling of Warp Knitted Structures. *Textile Research Journal* **72**, 266-272

[6] Kyosev Y, Renkens W 2007 3D-CAD für die Gestaltung von gewirkten Strukturen. In: 11. Chemnitzer Textiltagung, 24-25 Oktober 2007, pp. p 110-117

[7] Kyosev Y, Renkens W 2008 Prediction of Some Properties of Warp Knitted Structures Based on Numerical Simulation. In: DRAGCEVIC, Z. (ed.) Book of Proceedings for the 4th International Textile, Clothing & Design Conference "Magic world of textiles"
[8] Kyosev Y, Renkens W 2010 Modelling and visualization of knitted fabrics. In: Chen, X. (ed.) Modelling and predicting textile behaviour, pp. 225–262. Woodhead Publishing; In association with the Textile Institute; CRC Press, Cambridge, Boca Raton, Fla
[9] Kyosev Y, Renkens W 2010Numerical Simulation of the tension and compression behaviour of warp kintted reinforcements. In: Binetruy, C., Boussu, F. (eds.) Recent advances in textile composites. October 26-28, 2010, Lille Grand Palais, Lille, France, pp. 397-404. DEStech Publications, Lancaster, Pa
[10] Kyosev Y, Renkens W 2015 About the 3D Modelling of Jacquard Warp Knitted Structures, *J Fashion Technol Textile Eng* **s1**. doi: 10.4172/2329-9568.S1-003
[11] Honglian C, Mingqiao G, Gaoming J 2009 Three-Dimensional Simulation of Warp-knitted Fabric, *FIBRES & TEXTILES in Eastern Europe* **17**, 66–69
[12] Li X, Jiang G, Ma P 2015 Computer-Aided Design Method Of Warp-Knitted Jacquard Spacer Fabrics. *Autex Research Journal* **16**, 51-56. doi: 10.1515/aut-2015-0027
[13] Spencer D 2001 *Knitting technology*. Woodhead Publishing Limited, 2001
[14] Mista K 1996 Warp knitting machine with Jacquard-Control Patent EP 0583631 B1

Possible Applications of 3D Printing Technology on Textile Substrates

M Korger, J Bergschneider, M Lutz, B Mahltig, K Finsterbusch and M Rabe
Niederrhein University of Applied Sciences, Research Institute for Textile and Clothing (FTB), Mönchengladbach, Germany

E-mail: michael.korger@hs-niederrhein.de

Abstract. 3D printing is a rapidly emerging additive manufacturing technology which can offer cost efficiency and flexibility in product development and production. In textile production 3D printing can also serve as an add-on process to apply 3D structures on textiles. In this study the low-cost fused deposition modeling (FDM) technique was applied using different thermoplastic printing materials available on the market with focus on flexible filaments such as thermoplastic elastomers (TPE) or Soft PLA. Since a good adhesion and stability of the 3D printed structures on textiles are essential, separation force and abrasion resistance tests were conducted with different kinds of printed woven fabrics demonstrating that a sufficient adhesion can be achieved. The main influencing factor can be attributed to the topography of the textile surface affected by the weave, roughness and hairiness offering form-locking connections followed by the wettability of the textile surface by the molten polymer, which depends on the textile surface energy and can be specifically controlled by washing (desizing), finishing or plasma treatment of the textile before the print. These basic adhesion mechanisms can also be considered crucial for 3D printing on knitwear.

1. Introduction

3D printing is regarded as one of the key technologies in future production processes. In the field of additive manufacturing 3D objects are produced in layers using starting materials from liquid to solid and corresponding technological principles of model building named, e.g., stereolithography (SLA), selective laser sintering (SLS) or fused deposition modelling (FDM) [1].

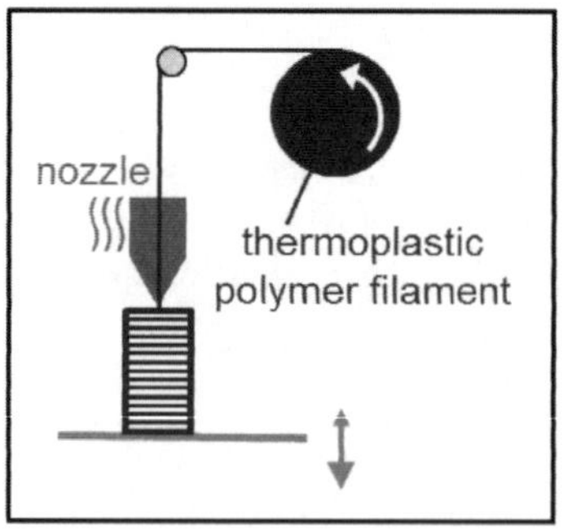

Figure 1. Schematic depiction of fused deposition modelling (FDM) technology.

The latter one is of special interest for SMEs due to its relatively low initial costs and easy operation compared to other 3D printing techniques. At FDM a wire-shaped thermoplastic filament is molten in an extruder nozzle and deposited in layers on a heatable printing bed, which is lowered according to the selected layer thickness each time one layer was printed. In this way the 3D element according to a previously designed CAD model is built up step by step (Fig. 1).

Combining FDM printing of thermoplastic polymers with textiles, new possibilities of individual product and process design arise. Previous investigations focussed on embedding fibrous materials [2] or woven fabrics [3] in 3D printed forms or printing on textile net structures creating new multicomponent textiles [4]. FDM printing on textile substrates, regarded as new kind of textile functionalization, offers production of individualized or customized products by freedom of shape and design concerning the fields of functional textiles as well as apparel. In addition to textile applications like buttons, accessories or 3D structured elements for home textiles (roller blinds, curtains) the possibilities to integrate a local reinforcement and protection function for sportswear as well as protection clothes can be considered as application fields.

While the adhesion characteristics of only FDM printed hard and stiff materials like acrylonitrile butadiene styrene (ABS), polylactic acid (PLA) and polyamide (PA) on different types of fabrics have recently been under investigation [5, 6], the current article concentrates on the flexible printing materials Soft PLA and a thermoplastic elastomer (TPE). As a result the influence of different chemical and physical surface properties of the textiles on the adhesion of the prints is depicted.

2. Experimental

3D printing experiments were performed using the FDM printer X400 manufactured by German RepRap GmbH. The selected nozzle diameter was 0.5 mm processing filaments with 3 mm diameter.

When testing the adhesion strength rectangles were created as CAD models, exported as stl files and imported in the Repetier-Host software for slicing. Rectangle dimensions for the prints usually were 150 mm x 50 mm. The layer thickness was set to 0.3 mm. Two layers were printed selecting a rectilinear infill pattern and a fill angle of 45°.

Soft PLA (shore hardness ≈ 43 D, purchased by German RepRap) and NinjaFlex (TPE filament, shore hardness ≈ 33 D) were used as printing filaments. The temperatures of the nozzle / printing bed were set to 210 °C / 70 °C (Soft PLA) and 220 °C / 35 °C (NinjaFlex). When printing the first layer it was ensured that the nozzle moves in close contact to the textile surface.

Adhesion tests were performed according to DIN 53530 (test for separating layers of laminated woven fabrics) using a Zwick Roell testing device. For evaluation the average peak values of the tensile forces in the recorded stress-strain-diagrams were given following the standard specifications.

Washing and desizing of textile samples were performed 45 min at 60 °C adding the enzyme amylase and the washing agent Kieralon CD (BASF).

Low pressure plasma treatments (50 Pa, 2.45 GHz microwave discharge) of textile samples were performed under carbon dioxide, argon and nitrogen atmosphere. The parameter settings are as following: 600 W (power), 250 sccm (gas flow), 30 s (duration of treatment).

3. Results and Discussion

Figure 2 shows maximum adhesion forces for both flexible Soft PLA and TPE filaments printed on different woven fabrics which can be mainly categorized by the textile material, weave and thickness. Obviously, the adhesion forces are more dependent on the textile substrate than on the printing material. A better adhesion can be obtained if the textile surface is roughened or hairy as shown for the rough polyester (PES), cotton (CO) and wool (WO) samples. Additionally, in case of the thicker fabrics also better adhesion forces can be acquired. These good adhesion results can be attributed to the provided form-locking connections of the printed polymer to the fibres on top of the textile as well as inside the textile structure, which should offer sufficiently open areas for the molten polymer to penetrate inside.

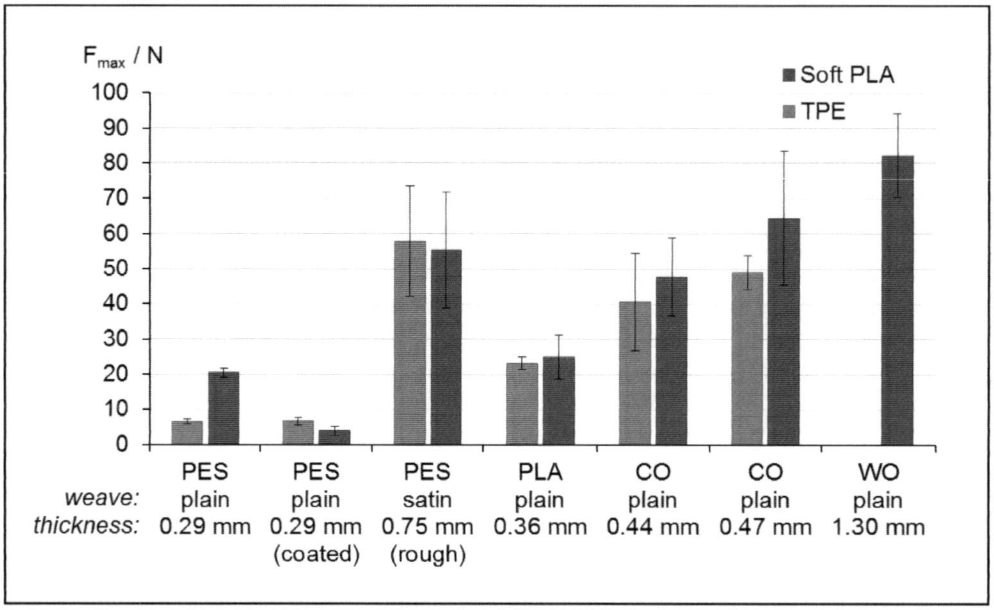

Figure 2. Maximum adhesion forces F_{max} of FDM prints (350 mm x 50 mm) on different woven fabrics – polyester PES, polylactic acid PLA, cotton CO and wool WO.

Apart from structural and topographical properties of the textile, the influence of the chemical properties of the textile surface providing, e.g., hydrophilicity or hydrophobicity on the adhesion strength was investigated. Figure 3 depicts the adhesion strengths of printed Soft PLA on different cotton and polyester fabrics for technical textile applications with similar thickness (0.21-0.33 mm) either unwashed or washed before the FDM printing process.

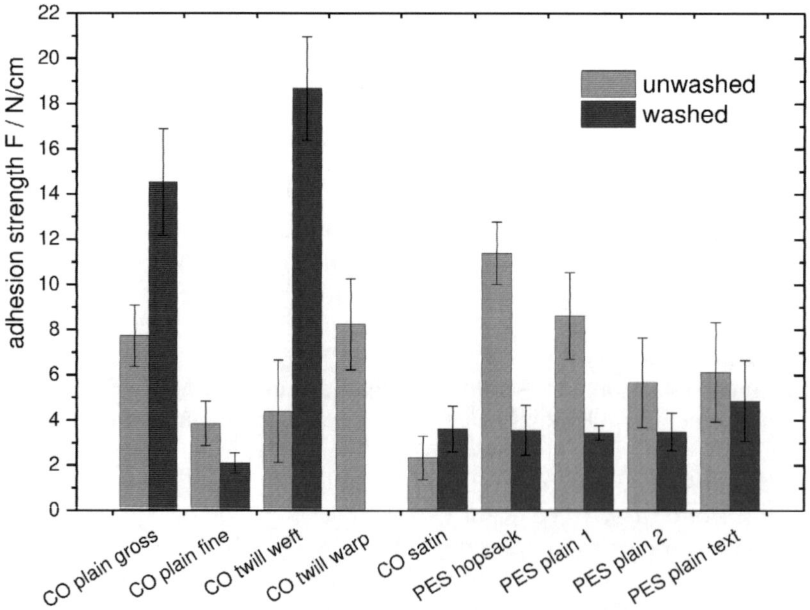

Figure 3. Adhesion strengths of Soft PLA prints on different unwashed and washed cotton CO and polyester PES fabrics.

During washing cycles, sizes and chemicals used for textile finishing should be removed. Looking at the cotton CO samples first, the determined values for the adhesion strength vary widely due to different textile surface structures. Except for CO plain fine, after washing and desizing the adhesion strength increases. For CO twill the adhesion strength even gets four times higher than without washing, so that it becomes impossible to remove the print by hand without fabric destruction. As proven by water droplet test, after washing the CO surface becomes hydrophilic owing to removal of waxes or lubricants. As a result the wettability of the textile surface and therewith the textile surface energy is enhanced.

In contrast to CO fabrics, washing treatment of PES fabrics led to hydrophobization of the PES surface, so that surface energy and wettability were reduced. This resulted in an adhesion strength reduction to similar values of around 4 N/cm. For this, it can be assumed that the formerly applied hydrophilic finish acts as a penetration and adhesion promoter.

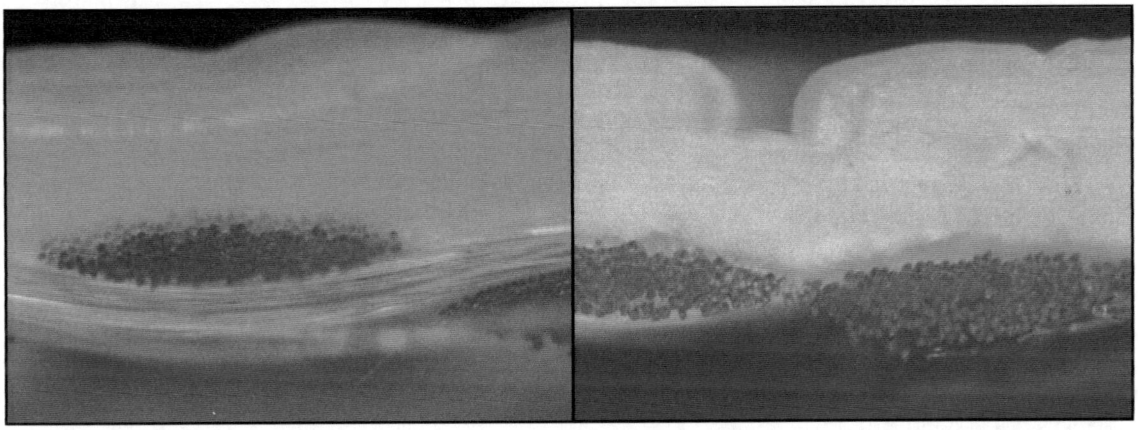

Figure 4. Cross section light microscope images of polyester PES samples without (left panel) and with argon plasma treatment (right panel) printed with Soft PLA on the top.

Low pressure plasma treatment of the polyester fabrics before the print led to different effects: Both argon (Ar) and nitrogen (N_2) plasma treatment drastically reduced the adhesion strengths for printed Soft PLA. Apparently, this kind of treatment only had a cleaning effect on the textile surface removing the hydrophilic finish and no polar groups were created. The resulting bad adhesion strength values of below 2 N/cm can be visualized by a cross-cut microscope image of the print revealing an air space at the interface between textile and printed polymer (Figure 4, right panel).

In case of carbon dioxide (CO_2) plasma treatment of the polyester fabric made of texturized yarns the hydrophilicity and wetting behaviour could be enhanced leading to almost two times higher adhesion strength for Soft PLA.

Using a Martindale testing device, various abrasion resistance tests with textile cotton and polyester samples printed with small squares of 2-3 layers of Soft PLA demonstrated, that the textile surface can be well protected against abrasion by FDM printed structures.

4. Conclusion

Deposition of polymers onto textiles using FDM printing technology can contribute to new textile applications and individual manufacturing of functional products. In this study the influence of textile surface properties on the adhesion strength of printed flexible polymers was examined considering mechanical / physical and chemical adhesion mechanisms. Basically, FDM prints on cotton and polyester textiles can result in great adhesion properties. The obtained adhesion strength is decisively influenced by the form-locking connections of the molten polymer with the textile substrate. These connections can occur on the textile surface (if roughened and hairy) and in the textile structure itself corresponding to the type of yarns and weave. Another important factor which affects the adhesion

IOP Conf. Series: Materials Science and Engineering **141** (2016) 012011 doi:10.1088/1757-899X/141/1/012011

strength is the wettability of the textile surface and so the textile surface energy. This is or can be specifically controlled by washing (desizing), finishing or plasma treatment of the textile before the print.

These basic principles resulting in good adhesion properties should also be valid for corresponding 3D printing processes on knitted structures, which will be the topic of further investigations.

5. References

[1] Chua C K, Leong K F and Lim C S 2003 *Rapid Prototyping: Principles and Applications, 2nd Edition* (Singapore: World Scientific Publishing Co. Pte. Ltd.)

[2] Richter C, Schmülling S, Ehrmann A, Finsterbusch K 2015 FDM printing of 3D forms with embedded fibrous materials *Design, Manufacturing and Mechatronics* 961-969

[3] Julius A, Lutz M, Finsterbusch K, Ehrmann A 2016 Integration of woven fabrics in 3D printed elements to enhance the mechanical properties *Technical Textiles* **59** E72-E73

[4] Sabantina L, Kinzel F, Ehrmann A, Finsterbusch K 2015 Combining 3D printed forms with textile structures – mechanical and geometrical properties of multi-material systems *IOP Conference Series: Mat. Sci. Eng.* **87** 012005

[5] Pei E, Shen J, Watling J 2015 Direct 3D printing of polymers onto textiles: experimental studies and applications *Rapid Prototyping Journal* **21/5** 556-571

[6] Grimmelsmann N, Lutz M, Korger M, Meissner H, Ehrmann A 2015 Adhesion of 3D printed material on textile substrates *submitted*

Acknowledgments

This work was supported by funding of the Ministry of Innovation, Science and Research of North Rhine-Westphalia.

Ministerium für Innovation,
Wissenschaft und Forschung
des Landes Nordrhein-Westfalen

IOP Conf. Series: Materials Science and Engineering **141** (2016) 012012 doi:10.1088/1757-899X/141/1/012012

Design methodology of the strength properties of medical knitted meshes

Z Mikołajczyk, A Walkowska
Lodz University of Technology / Department of Knitting Technology,
Zeromskiego st. 116, 90-924 Lodz, Poland

aleksandraannaw@gmail.com

Abstract. One of the most important utility properties of medical knitted meshes intended for hernia and urological treatment is their bidirectional strength along the courses and wales. The value of this parameter, expected by the manufacturers and surgeons, is estimated at 100 N per 5 cm of the sample width. The most frequently, these meshes are produced on the basis of single- or double-guide stitches. They are made of polypropylene and polyester monofilament yarns with the diameter in the range from 0.6 to 1.2 mm, characterized by a high medical purity. The aim of the study was to develop the design methodology of meshes strength based on the geometrical construction of the stitch and strength of yarn. In the environment of the ProCAD warpknit 5 software the simulated stretching process of meshes together with an analysis of their geometry changes was carried out. Simulations were made for four selected representative stitches. Both on a built, unique measuring position and on the tensile testing machine the real parameters of the loops geometry of meshes were measured. Model of mechanical stretching of warp-knitted meshes along the courses and wales was developed. The thesis argument was made, that the force that breaks the loop of warp-knitted fabric is the lowest value of breaking forces of loop link yarns or yarns that create straight sections of loop. This thesis was associate with the theory of strength that uses the "the weakest link concept". Experimental verification of model was carried out for the basic structure of the single-guide mesh. It has been shown that the real, relative strength of the mesh related to one course is equal to the strength of the yarn breakage in a loop, while the strength along the wales is close to breaking strength of a single yarn. In relation to the specific construction of the medical mesh, based on the knowledge of the density of the loops structure, the a-jour mesh geometry and the yarns strength, it is possible, with high accuracy reaching up to 95 – 87 %, to determine the actual bidirectional strength of medical meshes.

1. Introduction

Warp knit fabrics are widely used, among others, in clothing, underwear production and hosiery, decorating industry, upholstery as technical products, including medical devices. The last group can include knitted medical mesh, among others, **hernia mesh** [1, 2, 3, 4]. Hernia surgeries are one of the most commonly performed medical procedures. It is worth noting that ignoring hernia may result in serious consequences, and even death.

Hernia (Lat. Hernia) is an undesirable exit of the whole or part of an organ or bulging of tissue beyond its anatomical location [6, 7, 8, 9]. In 1984, Irving Lichtenstein performed the first surgery using hernia mesh (Fig.1).

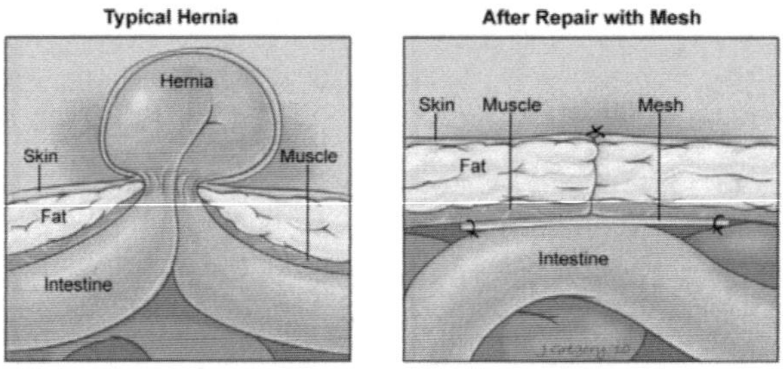

Fig.1. View of the body structures before and after surgery using hernia mesh [5]

The basic performance parameter of knitted hernia mesh is its bi-directional tensile strength along courses and wales. In the process of designing knitted fabric, we should also take into account the need to reduce the weight of an implant, ensure a high porosity of the resulting structure and its elasticity as well as its biocompatibility with the body. The expected value of the strength of the hernia mesh is estimated by manufacturers and doctors at the level of 20÷30 cN/11 cm of the width of a sample [10]. Currently, the process of designing hernia mesh is a trial-and-error process, since structural parameters and strength are tested empirically, after the fabric has been manufactured. There is no logical and engineering methodology of designing bi- or multi-directional strength of medical mesh. Therefore, it was advisable to undertake research to develop the methodology to design strength properties of medical mesh in terms of the geometric structure of the weave and yarn properties.

2. The mechanical model of the stretched warp knit fabric.

An analysis of the forces acting in the elements of the structure of the mesh during the stretch of the fabric in static conditions has been performed. Considerations refer to stretching along courses and wales. Across the width S of the fabric stretched by force F there is n wales and m courses, each fabric is then stretched with a force $\frac{F}{n}, \frac{F}{m}$. A serial arrangement of the structural elements of the fabric results in the stretching process which can be reduced to one mesh. This allows us to specify an absolute strength for the whole width of the sample, and a relative strength for a course or a wale of the mesh. For more complex structures the strength is referred to a repetitive element, for example, a slot (intersection of meshes) or a unit of the length and the width. Currently, manufacturers usually provide the absolute strength of the mesh in Newton, defined for a sample with the width equal to 5 cm. Looking at the problem from the viewpoint of the structure of the fabric, it is justified to use the concept of relative strength. This enables the design of the strength for similar structures of a greater or lesser density and of a comparable structure of repetitive elements of the knitted fabric.

In Figure 2 a model of the knitted fabric stretched along courses and along wales has been presented. A sample of the width S is gripped between two jaws, the upper one moves sliding up to the breaking point. The distribution of forces in the serially arranged mesh while stretching along the courses and wales has been presented in Fig. 3.

The following thesis has been put forward for research on modeling of the strength of knitted fabrics: The force breaking the mesh of a warp knit fabric is the lowest value of the force from among forces breaking the thread of the mesh chain or the force breaking thread of straight sections of the mesh.

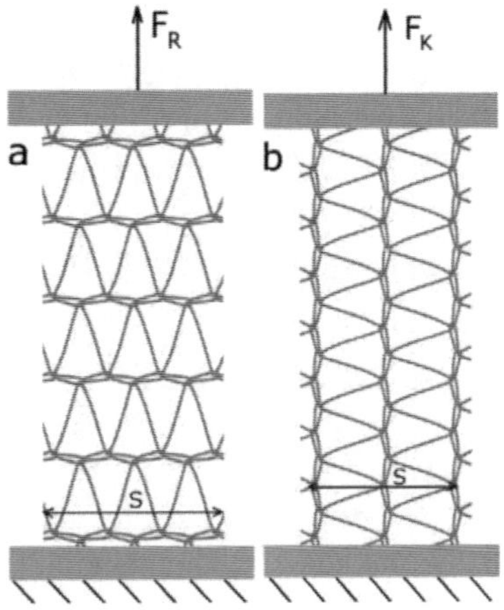

Fig. 2. The scheme of the stretched fabric
a) along courses b) along wales

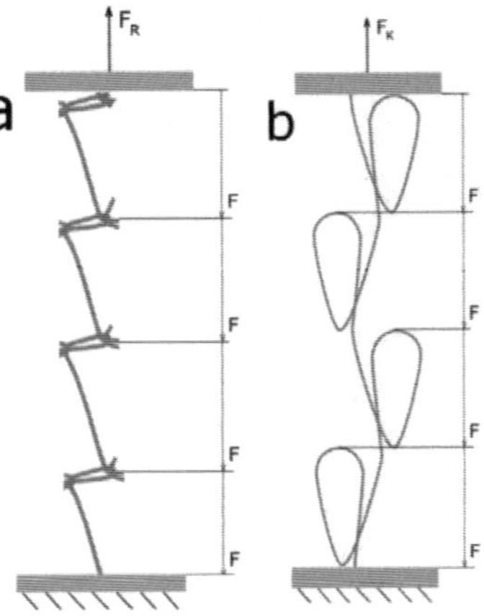

Fig. 3. The scheme of a stretched
a) course b) wale

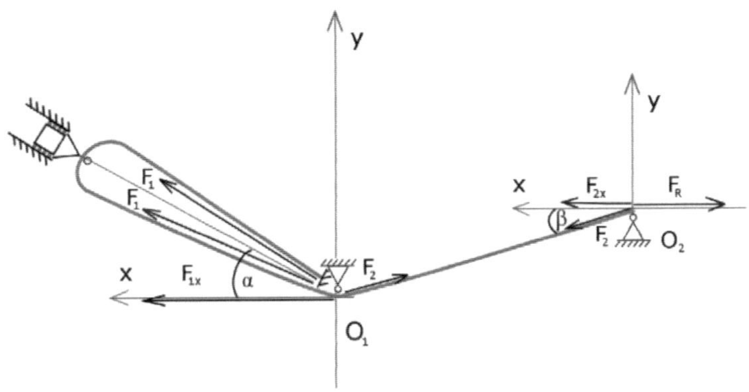

Fig. 4. A model of static stretching of the mesh in warp knit fabric along courses

F_1- forces acting in straight sections of the loop of the mesh
F_2- a force acting in the connector of the mesh
$F_{1,2x}$- a component of forces acting in the threads of the loop in the direction of stretching the mesh (axis OX)
F_R- a force stretching one course
α- an inclination angle of an axis of the loop from axis OY
β- an inclination angle of the connector from axis OY

This thesis, which is the foundation for the theoretical analysis of the strength of knitted fabric, refers to the theory of the strength and fatigue life using the concept of "the weakest link" (the weakest link concept) [6]. The defect of an elementary link in the described concepts of probability is understood as any object in the fabric, which generates heterogeneity of the fabric structure and accelerates the process of degradation. Therefore, a criterion of minimum force for all considerations concerning the determination of the mechanical strength of the mesh has been adopted:

$$Fz = F\ min.\ (F\ R\ min.\ i\ F\ K\ min.),$$

where: Fz– the force breaking the mesh of the fabric, FR min. – the minimum force breaking the threads in the mesh of the fabric stretched along the courses, FK min. – the minimum force breaking the threads in the mesh of the fabric stretched along the wales.
The correctness of the presented thesis will be the answer to the question: "Which of the forces acting on the threads of the mesh with a specific strength of this thread is the lowest force that causes the damage to the mesh?"

2.1 Stretching the fabric along the courses.

An analysis of the forces acting in the mesh of a warp knit fabric has been performed subject to tensile force F_R. The distribution of forces in the components of the mesh, i.e. in the connector and straight sections of the loops of the mesh has been shown in Fig. 4.

Equilibrium conditions according to the reference system XYO_1:

$$-F_2 \cos\beta + 2F_1 \cos\alpha = 0 \quad (2.1\text{-}1)$$

If: $$F_R = F_2 \cos\beta \quad (2.1\text{-}2)$$

then: $$F_R = 2F_1 \cos\alpha \quad (2.1\text{-}3)$$

Equilibrium conditions according to the reference system XYO_2:

$$F_{2x} = F_2 \cdot \cos\beta \quad (2.1\text{-}4)$$

$$F_2 \cos\beta - F_R = 0 \quad (2.1\text{-}5)$$

Breaking forces:

$$F_{R1} = 2F_{ZN} \cos\alpha \quad (2.1\text{-}6)$$

$$F_{R2} = F_{ZN} \cos\beta \quad (2.1\text{-}7)$$

The force which will have a lower value according to formulas 2.1-6 and 2.1-7 is adopted as the force breaking the mesh while stretching the fabric along the courses:

$$F_{ZR} = F_{Rmin.}$$

2.2 Stretching the fabric along the wales.

An analysis of the forces acting in the mesh of a warp knit fabric subject to tensile force FK has been performed. Three variants of the geometry of the forces in the elements of the structure of the mesh have been adopted, i.e. in the connector and in straight sections of the mesh for the following relations of angles γ and δ:

I. where $\gamma = \delta$

II. where $\gamma > \delta$

III. where $\gamma < \delta$.

To solve the equations of equilibrium of forces in the elements of the mesh to force F_K in the case of the relations of angles $\gamma>\gamma$ and $\delta<\delta$, the transformation of the reference system OXY has been adopted, so that one of the components of the force F_K is equal to zero,
i.e. F_{1y}` = 0 then F_K = f (F_2 and trigonometric functions of angles γ and δ)
and F_{2y}` = 0 then F_K = f (F_1 and trigonometric functions of angles γ and δ)
The smallest value must be selected from the calculated values, which will determine the force breaking the wale:

$$F_{ZK} = F_{Kmin.}$$

Fig. 5 shows a generalized model of stretching the fabric along the wales taking into account the movement of threads between the loops and a connector of the mesh. In the course of further discussion this phenomenon has not been taken into account.

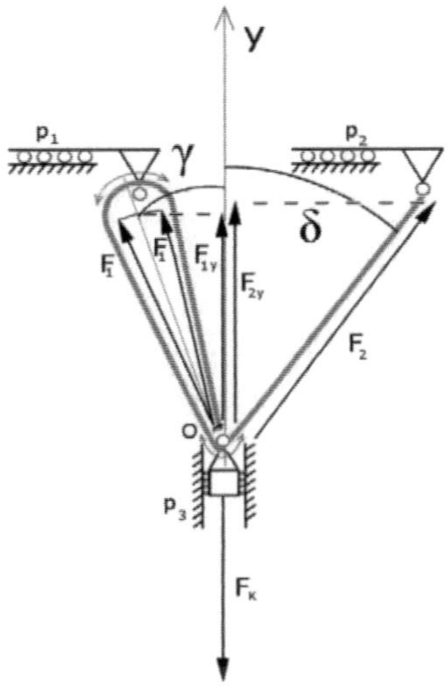

F_1- forces acting in straight sections of the loop of the mesh
F_2- a force acting in the connector of the mesh
$F_{1,2y}$- a component of forces acting in the threads of the loop in the direction of stretching the mesh
F_K- a force stretching (breaking) one wale
γ- an inclination angle of an axis of the loop from axis OY
δ- an inclination angle of the connector from axis OY
p_1, p_2, p_3 – expected displacement of threads

Fig. 5. A model of static stretching of the warp knit fabric along the wales

Table 1. Models used to determine the force breaking the mesh while stretching along the wales.

Relation of angles $\gamma = \delta$	$F_{K1} = 3F_{ZN}\cos\gamma$	
	Relation of angles $\gamma>\delta$	**Relation of angles $\gamma<\delta$**
Elimination of forces acting in the loop of the mesh		
Force F_2 to the left of axis y\`	$F_{K2} = \dfrac{F_{ZN}\cos\left[\frac{\pi}{2} - (\gamma+\delta)\right]}{\cos\left(\frac{\pi}{2} - \gamma\right)}$	$F_{K8} = \dfrac{F_{ZN}\cos\left[\frac{\pi}{2} - (\gamma+\delta)\right]}{\cos\left(\frac{\pi}{2} - \gamma\right)}$
Force F_2 to the right of axis y\`	$F_{K3} = \dfrac{F_{ZN}\cos\left[\delta - \left(\frac{\pi}{2} - \gamma\right)\right]}{\cos\left(\frac{\pi}{2} - \gamma\right)}$	$F_{K9} = \dfrac{F_{ZN}\cos\left[\delta - \left(\frac{\pi}{2} - \gamma\right)\right]}{\cos\left(\frac{\pi}{2} - \gamma\right)}$
Force F_2 on axis y\`	$F_{K4} = \dfrac{F_{ZN}}{\cos\left(\frac{\pi}{2} - \gamma\right)}$	$F_{K10} = \dfrac{F_{ZN}}{\cos\left(\frac{\pi}{2} - \gamma\right)}$
Elimination of a force acting in the connector of the mesh		
Force F_1 to the left of axis y\`	$F_{K5} = \dfrac{2F_{ZN}\cos(\delta - \gamma)}{\cos\left(\frac{\pi}{2} - \delta\right)}$	$F_{K11} = \dfrac{2F_{ZN}\cos\left[\gamma - \left(\frac{\pi}{2} - \delta\right)\right]}{\cos\left(\frac{\pi}{2} - \delta\right)}$
Force F_1 to the right of axis y\`	$F_{K6} = \dfrac{2F_{ZN}\cos\left[\frac{\pi}{2} - (\gamma+\delta)\right]}{\cos\left(\frac{\pi}{2} - \delta\right)}$	$F_{K12} = \dfrac{2F_{ZN}\cos\left[\left(\frac{\pi}{2} - \delta\right) - \gamma\right]}{\cos\left(\frac{\pi}{2} - \delta\right)}$
Force F_1 on axis y\`\`	$F_{K7} = \dfrac{2F_{ZN}}{\cos\left(\frac{\pi}{2} - \delta\right)}$	$F_{K13} = = \dfrac{2F_{ZN}}{\cos\left(\frac{\pi}{2} - \delta\right)}$

A collective presentation of equations to calculate breaking force FK of the stretched knitted fabric along the wales for three variants of relations of angles γ and δ and three cases of the location of forces F_1 and F_2 to the axis of OY` have been shown in Table1.
The listed relations to determine the breaking force are a function of the strength of the thread and geometric parameters of the mesh i.e. FK = f(FZN ; γ andδ). To model the strength of the knitted mesh the data is needed regarding the strength parameters and the geometry parameters of the structure of the knit in the process of stretching.

3. An analysis of the structure of the fabric while stretching.
An analysis of the geometry of the structure of a warp knit fabric focused on the determination of the following parameters:

- inclination angle α, γ of the loop of the mesh to axis OX and OY,
- inclination angle β,σ of the connector to axis OX and OY,
- width of the wale, A
- height of the course, B
- width of the loop of the mesh, c
- height of the loop of the mesh, d
- size of satin area, Pp (only in Pro Cad program for loosely structured variants)

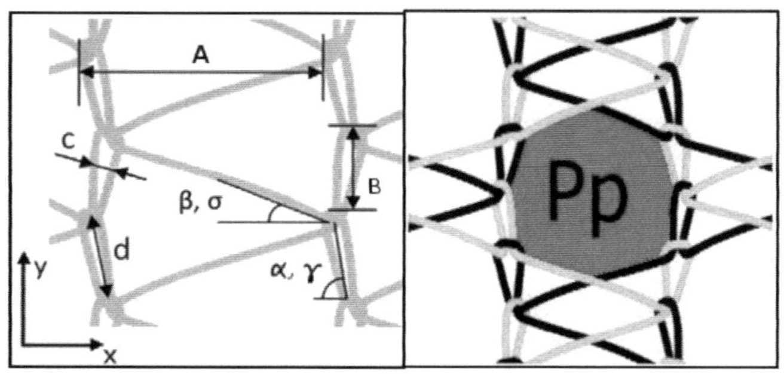

Fig. 6. Analyzed structures of knitted mesh

The analysis was carried out on the basis of images of the stretched fabric on the tensile tester.
For the purposes of empirical research six variants of the mesh weaves have been designed: symmetrical double course satin with cloth structure and mixed meshes for threadings 1/3, 2/2, 3/1, 1/1 (with shifted and identical threadings) and a weave of the cloth of threading 1/1.
The knitted fabric was manufactured from polypropylene monofilament yarn having a diameter of 0.08 mm, on a warp HKS3 produced by K. Mayer, with a needling number E28.Photos have been taken of stretched fabrics on the tensile testing machine in two perpendicular directions (Fig. 7).

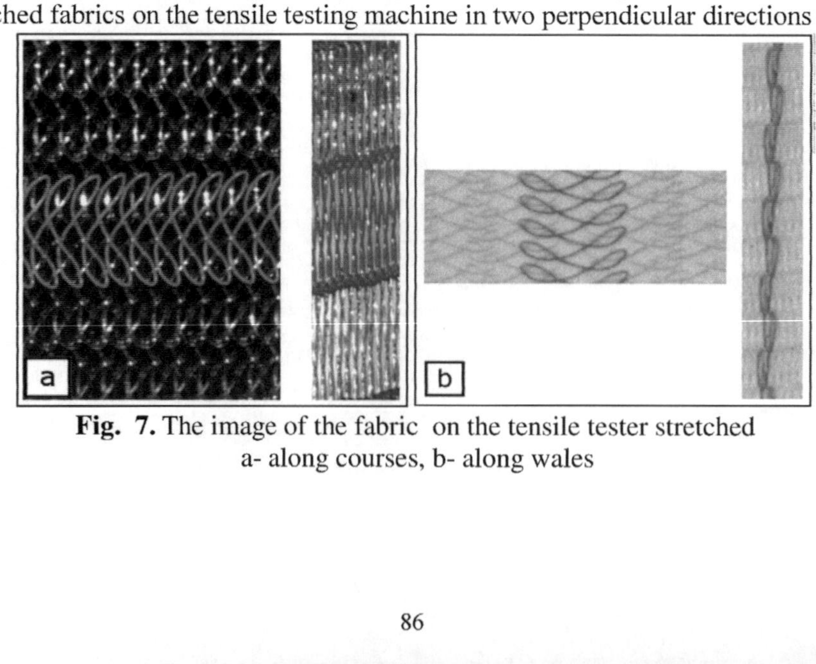

Fig. 7. The image of the fabric on the tensile tester stretched
a- along courses, b- along wales

On the basis of the photographs taken with the program GIMP basic geometrical parameters of the structure of the knitted fabric have been read. The results are shown in Table 2.
Stretching both along courses and wales causes a significant rapprochement of straight sections of the mesh and connectors to axis OY (a longitudinal axis of the stretched fabric), causing that inclination angles γ and σ at the breaking point are close to zero. The values of angles obtained in this method were used in the calculation of the critical values of forces breaking the fabric mesh.

4. The experimental analysis of the strength of warp knit fabrics

4.1 Testing the strength of yarn
The strength of single yarn and yarn in the loop has been tested. Polypropylene yarn with a linear density of 4.6 tex and a diameter of 0.08 mm was subject to the test. The test was conducted on the device manufactured by Hounsfield, setting the speed of the upper beam movement at 100 mm/min. The length in jaws was 500 mm. The study shows that the breaking strength of the monofilament in the loop is 3.72 N with elongation at the breaking point of 19.62 %, so per thread the force is 1.86 N. The average strength of a single thread stretch is 2.16 N with elongation $\varepsilon = 25.94$ %.

Table 2. The values of the basic parameters of the structure of the mesh while stretching along courses and wales

STRETCHING ON TENSILE TESTER ALONG COURSES		**STRETCHING ON TENSILE TESTER ALONG WALES B**	
Elongation: initial	maximum	initial	maximum
inclination angle of the connector to axis OX, degrees			
59.22	85.91	21.25	85.25
inclination angle of the loop to axis OX. degrees			
66.85	88.28	17.36	85.68
inclination angle of the connector to axis OY. degrees			
30.78	4.09	68.75	4.75
inclination angle of the loop to axis OY. degrees			
23.15	1.72	72.64	4.32
width of two loops and the connector . mm			
4.03	4.64	4.32	0.72
width of a wale. mm			
2.262	too close arrangement of elements of the mesh	2.41	too close arrangement of elements of the mesh
height of a course. mm			
0.41	too close arrangement of elements of the mesh	0.89	too close arrangement of elements of the mesh
width of a loop. mm			
0.42	too close arrangement of elements of the mesh	0.54	0.18
height of a loop. mm			
1.47	1.15	1.54	1.54

4.2 Testing the strength of the knitted fabric

The study was conducted on two strength testing machines: HOUNFIELD and Instron 3345. The samples used for testing the strength of knitted fabrics have dimensions 5x20 cm, the distance in the jaws of the tensile machine is 10cm. Knitted mesh with the weave of satin and cloth were subject to the test. Only the basic weave of cloth of threading 1/1 was further analyzed. The big problem for both tensile machines while stretching was breaking at the jaws.

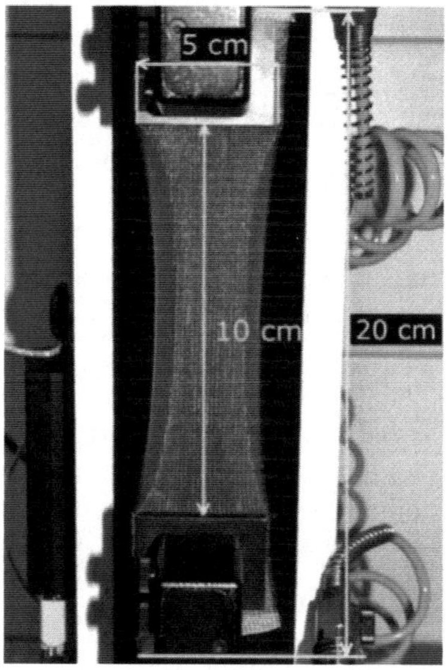

Fig. 8. View of knitted fabric while stretching

The measurement results of stretched knitted fabrics:

The average absolute strength of the fabric while stretching along wcourses was 150.25 N, elongation at the breaking point of 108.1 %. The absolute strength along the wales: 89.78 N, elongation at the breaking point of 316.9 %. Referring the above values to a single course and single wale we receive, respectively:

the relative strength along the courses:1.67 N / 1 course,

the relative strength along the wales: 2.24 N / 1 wale.

5. Verification of methodology of designing the strength of knitted medical mesh

Calculations have been made of the theoretical strength of the knitted mesh with a cloth weave, based on the models identified in Chapter 2. These values were determined depending on the inclination angles of the elements of the mesh, i.e., the connector and straight sections of the loop to axis OY and the strength of single yarn and yarn in the loop. The calculation algorithm was created in Excel. It consists of three blocks:

- input data: the strength of thread F_{ZN}, the angles α, β, γ, δ
- computing, which recorded fifteen formulas determining the breaking forces of knitted fabric
- results of the calculation: the value of the minimum force of all calculated values in the calculation block ($F_{Rmin.}$ i$F_{Kmin.}$).

5.1 The computing strength of the knitted fabric stretched along the courses

Since a criterion $F_{ZR} = F_{Rmin.}$, was established as the force breaking the mesh, a smaller force is accepted from the obtained calculations, i.e. **F_{R2} which equals to 1.86 N/wale.**

Also, calculations of the strength of yarn broken individually were performed giving the result F_{ZR}=2.15 N/1course. The relative strength of the fabric stretched along the courses, referred to one mesh can be determined by means of the strength of the yarn in the loop. It equals 1.86 N. The absolute strength of the fabric is then determined as the product of the force and the number of courses in the sample.

5.2 The computing strength of the knitted fabric stretched along the wales

To determine a theoretical force breaking the mesh of the fabric stretched along the wales, it was necessary to determine the relationship between the angles γ and δ, and perform appropriate calculations. In the studied weave of cloth a relation $\gamma<\delta$ occurs between the angles, so the formulas FK8-FK13 were used. According to the criterion of a minimum force, the smallest force FKmin is considered to be a value breaking the mesh.

After rounding measurements of angles γ and δ to the nearest unit, 4 degrees were obtained for both angles. Thus it is reasonable to use the following formula:

$$F_K = 3F_{ZN}\cos\gamma = 3\cdot 2{,}16\ \mathrm{N}\cdot\cos 4^\circ = 3\cdot 2{,}16\ \mathrm{N}\cdot 0{,}99 = 6{,}42\ \mathrm{N/1course}$$

Table 3. Computing strengths

Determining the strength of the mesh while stretching along wales		
For mean values		
	value	unit
force breaking yarn in the loop	1.86000	N
average inclination angle of the loop to axis OY. γ	4.31800	degrees
γ. rad	0.07536	radians
average inclination angle of the connector to axis OY. δ	4.75200	degrees
force FK8. breaking a wale. 3.89434 N		
force FK9. breaking a wale. 3.89434 N		
force FK10. breaking a wale. 24.70382 N		
force FK11. breaking a wale. 7.07875 N		
force FK12. breaking a wale. 7.07875 N		
force FK13. breaking a wale 44.90422 N		
Minimal force determining the strength of warp knit fabric while stretching along wales		
ForceFKmin.	3.89434	N

This result differs significantly from the actual value of the force breaking one mesh, which is equal to 2.24 N / wale. The following is a calculation algorithm of forces along the wales for the case $\gamma<\delta$, adopting the force breaking the thread in the loop (Table 3).

The minimum value of FKmin=3.89 N/1wale was obtained. Calculations taking into account a force breaking a single yarn were performed. For this case, the theoretical strength of the mesh is FKmin.= 4.52 N/1wale.

From this comparison it follows that experimental and theoretical values of the strength along the courses are close, however, while stretching along wales they are significantly different. Noteworthy is a large convergence of the fabric strength to the strength of threads. Attention should also be given to the problems of measuring the angles necessary to determine the designed strength. This applies to the place of measurement of angles and a big discrepancy between the measurement results. It seems legitimate for practical reasons to estimate the relative strength of the fabric on the basis of the strength of the yarn used.

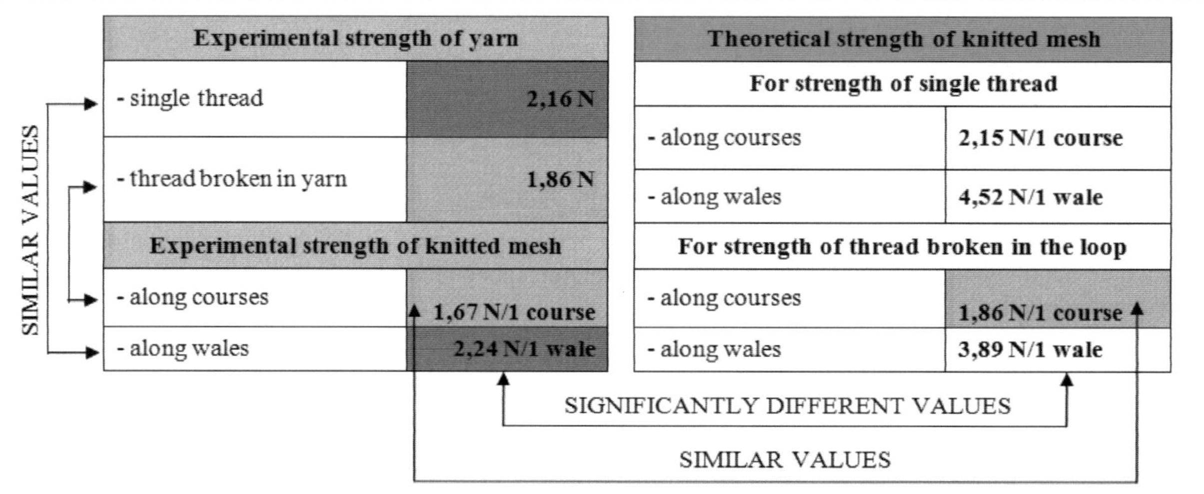

Fig. 9 Verification of simulations of the strength of knitted mesh

6. Conclusions

1. In the absence of logical and engineering method of designing single- or multi-directional strength of medical mesh, it was expedient to develop theoretical assumptions to identify strength properties of medical mesh in terms of the geometry of the weave and properties of yarn.
2. The mechanical model of the strength of knitted mesh was based on the thesis that a force breaking the mesh of a warp knit fabric is the lowest value of the force from among forces breaking the thread of the mesh chain or a force breaking the thread of straight sections of the mesh. This thesis refers to the theory of strength and fatigue life based on the probabilistic concept of the "weakest link".
3. On the basis of the presented models of two-way process of stretching fifteen calculation relations of a critical value of a breaking force were introduced. The strength of the mesh is a function of the strength of yarn and an inclination angle of the sections of the mesh to the axis of stretching.
4. Calculated values of the force breaking the mesh of knitted fabric are close to the experimental strength.
5. The actual strength of the cloth weave mesh related to one course is equal to the strength of yarn broken in the loop. The strength of the fabric stretched along the wales, however, is close to the force breaking a single yarn. The following statement can be used in a practical approach to estimate the approximate strength of knitted mesh.
6. The absolute strength breaking the warp knit fabric subject to stretching is directly proportional to the strength of yarn and closeness of courses and wales.

References

[1] Janczak D, Litarski A, Merenda M, Rać J, Wieraszko A, Litarski A 2011 Polimery w Medycynie, *Stosowanie siatek propylenowych w operacjach przepuklin brzusznych u chorych, operowanych z powodu tętniaka aorty brzusznej*,T. 41, Nr 3,

[2] Struszczyk MH, Olejnik M. 2010 Present and future demand for textile medical devices (in Polish). Technicznewyrobywłókiennicze **18**, 19-25

[3] Niekraszewicz A, Kucharska M, Wawro D, Struszczyk MH, Kopias K, Rogaczewska A. 2007 Development of a Manufacturing Method for Surgical Meshes Modified by Chitosan, *Fibres& Textiles in Eastern Europe* **15**, 105-109

[4] Struszczyk MH, Komisarczyk A, Krucińska I, Kowalski K, Kopias K.: Ultra-Light Knitted Structures for Application in Urologinecology and General Surgery – Optimization of Structure, *FiberMed11* 28-30 June 2011, Tampere, Finland, ISBN 978-952-15-2607-7

[5] Widok struktur ciała przed (a) i po operacji (b) z użyciem siatki [online]. Dostępny w Internecie: http://www.wehealny.org/services/bi_hernia/images/hernia.jpg

[6] Klosterhalfen B. et al. 2005 The lightweight and large porous mesh concept for hernia rep air, *Expert Rev. Devices* **2**, 1-15

[7] Struszczyk MH et al. 2005 Surgical Meshes – Physiological Aspects, MedTex, *Proceedings of the Vth International Scientific Conference*, ISBN 83-911012-3-1, 112

[8] Goldstein HS 1999 Selecting The Right Mesh, *Hernia* **3**, 23-26

[9] Struszczyk MH, Gutowska A, Kowalski K, Kopias K, B Pałys, Komisarczyk A, Krucińska I 2011 Ultra-Light Knitted Structures for Application in Urologinecology and General Surgery, – Optimization of Structure in the Aspect of Physical Parameters, *Fibres & Textiles in EasternEurope* **19**, 92-98

[10] Bilsel Y, Abci I 2012 The search for ideal hernia repair; mesh materials and types, International *Journal of Surgery* **10**, Istambuł, Turcja

The properties of weft knitted fabric medical and preventive treatment action using eco-raw materials

L Halavska, O Batrak
Kyiv National University of Technologies and Design, Ukraine

Presenting author e-mail: ludagal73@list.ru; corresponding author e-mail: sasha01021@gmail.com

Abstract. A new trend in the world is the clothing production using the new types of ecological raw materials application – milk, pineapple, coconut, hemp, banana, eucalyptus, clams, corn, bamboo, soya, nettle yarn. This makes it possible to create textile materials of new generation with unique antibacterial and antiseptic properties. Such materials have a positive preventive and sometimes therapeutic effect on people, and their health. Eco-raw materials clothing is able to protect the human body from the environment harmful effects: cold, heat, rain, dust, opportunely remove from underclothing layer the steam and gases, sweat; maintain in underclothing layer the necessary microclimate for normal organism functioning. Study of knitwear consumer properties, produced with eco-materials, is an urgent task of the world vector, directed on ecological environmental protection.

This paper presents the research results of hygroscopicity and capillarity weft knitted fabrics, what knitted from different types of eco-raw materials: bamboo yarn, yarn containing soybean and nettle yarn. Character of influence of the liquid raising level changes depending on the experiment time and the knitting structure is revealed.

1. Introduction

Bamboo tree easily grows by itself in areas with good ecology. It does not require any additional farm work, special watering. Bamboo does not require protection from mosquito as spraying the soil and the plant with chemical fertilizers. It has antibacterial properties, prevents increase of harmful microorganisms. Bamboo textile materials are 20 % more breathable than cotton ones and absorb water 60 % better. Such properties are available owing to the great many of natural pores in the fibre. Breathability and hygroscopicity are determined with the degree of porosity of the fibre. Bamboo fibres absorb and evaporate water from the whole surface. If required, bamboo may output excessive heat and vice versa keep the warm in cold season. Thus, moisture and temperature optimal for the body are provided. That's why, bamboo clothes prevent perspiration, the body becomes cool, and there is no unpleasant smell. Moreover, bamboo textile materials protect from UV radiation up to 98 %. It is an ideal material for summer clothing [2, 3].

Soya fibre as well as bamboo one is ecologically raw material. Soya fibre is produced with the latest technology based on processing of plant proteins from soya beans. As well as bamboo, soya beans easily grow and they are insect resistant. There is no need in large number of pesticides, fertilizers and genetically modified objects. Textile materials from soya fibre are very tactile, wearable and easy to look after. They do not require hot water in washing, linear dimensions are not changed in washing, and materials dry very fast. Clothes from soya fibre absorb and evaporate water in a moment [4].

Technology for production of small and great nettle appears not long ago. The nettle grows near the roads, on wastelands, as usual weeds. Certain technology enables us to process the nettle and produce textile materials for clothing. It is proved that nettle clothes helps the customer for certain

ailment: headache, pain in the joints. Clothes from nettle improve blood circulation and have other medical and recreational effect.
Nettle raw material is manufactured in a similar way to flax. Technology for production of nettle yarn is not complicated but rather time-consuming. Classic proceeding includes the following stages: soaking, drying, breaking, shaking, hackling and spinning of fibres [5].

The main factor for using eco-friendly materials in clothing is the fact that all of them are decayed in soil due to microorganisms and do not emit harmful substances to the environment. Processing of such types of fabric on knitting equipment is not well-studied. There is no information about influence of structural properties to physical properties, in particular to hygroscopicity and capillarity [4, 5, 6, 7].

2. Experimental

For the raw material to produce experimental models, the following types of yarn were taken: bamboo yarn (100 %) linear density 40x9 tex, yarn that contains soya bean (50 % soya and 50 % bio-cotton) linear density 25x8 tex and nettle yarn (100 %) that is a knitted ribbon made with rib knitting structure 1+1 - 355 tex. Experimental models are made on plane-knitting machine 'Brother' of the 5-th class with plain and rib knitting structures. Density levels of knitting are defined for the machine; it provides standard process of loop formation.
Hygroscopicity and capillarity of experimental models are determined according to standard method. It is commonly known that hygroscopicity of textual materials describes their ability to absorb and evaporate water and water vapour [8]. Experimental results are presented in histogram (Fig. 1).

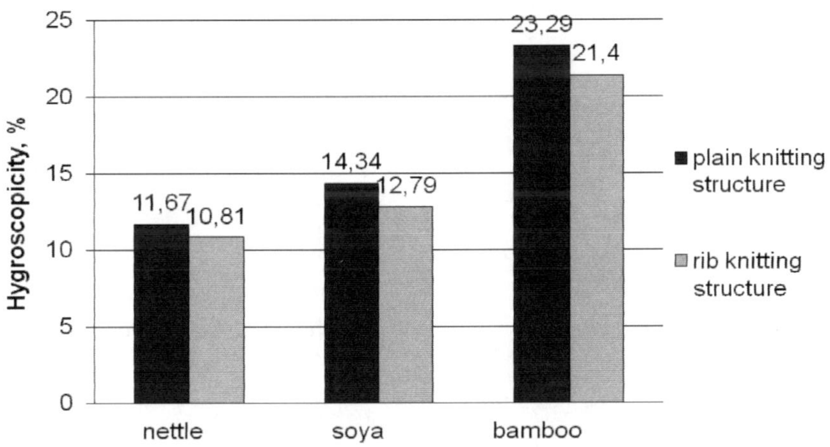

Fig. 1. **Hygroscopicity of experimental models**

As you see from histogram, bamboos knitted models have max hygroscopicity. For knitted fabric with plain structure this value is 23 %, with rib structure – 21 %. Nettle fabric has min hygroscopicity (plain structure – 12 %, rib structure – 11 %). During experimental survey it is discovered that hygroscopicity depends not only on raw material but also on peculiar features of fabric structure formation. Despite the fact that surface density of fabric models with rib structure is higher, these models illustrate lower level of hygroscopicity. This may be explained by peculiarities of structure formation. Knitted fabric with rib structure has two layers of stitches (purl stitch is located beyond the knit stitch because of elastic properties of yarn) while with plain structure – only one layer. Thus, vapour liquid does not completely penetrate to the internal layer of fabric with rib structure and, as a result, models have lower level of hygroscopicity.

Process of raising the liquid level along the course and wale for the experimental models with plain structure during 60 minute experiment is presented in Fig. 2 and Fig. 3, respectively, for the models

with rib structure 1+1 – in Fig. 4 and Fig. 5, respectively.
Generated diagrams enable us to find out the nature of raising the liquid level depending on duration of experiment. Appropriate equations of regression are obtained.

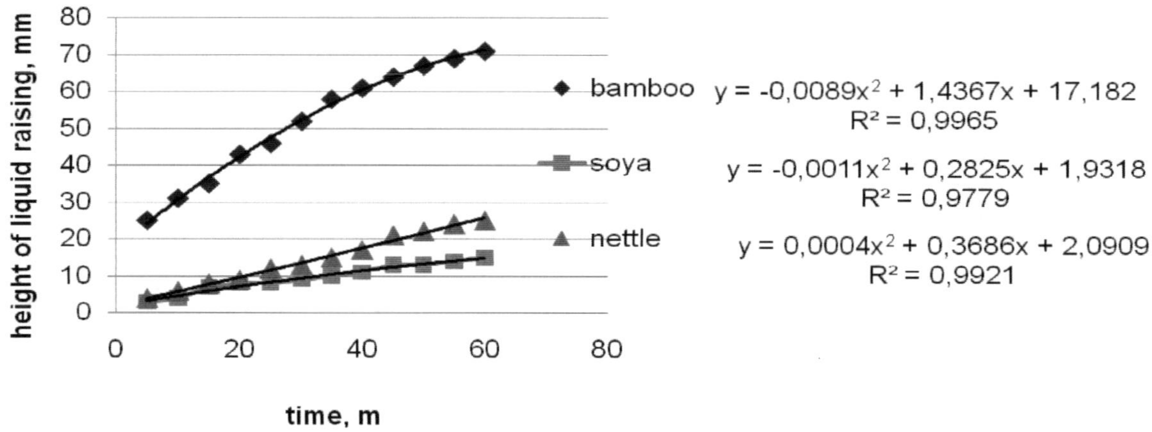

Fig. 2. Diagram of raising the liquid level of experimental models with plain structure along the wale.

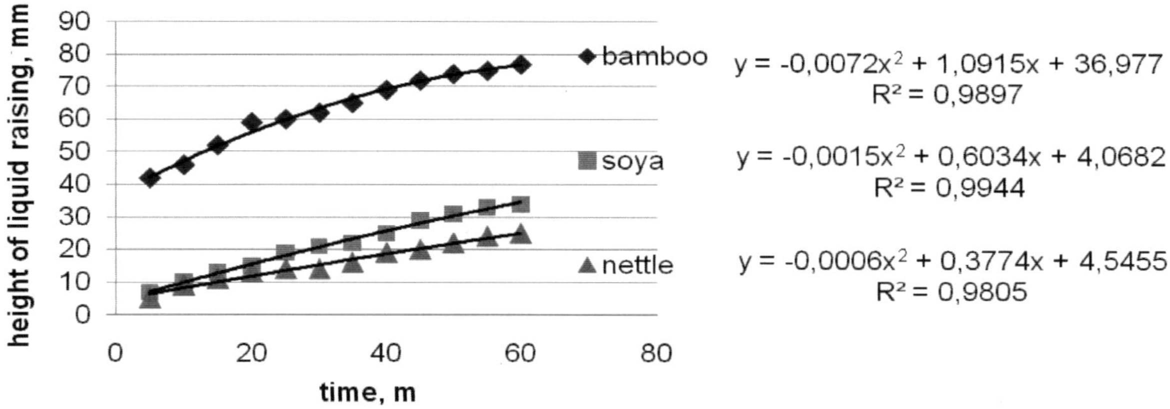

Fig. 3. Diagram of raising the liquid level of experimental models with plain structure along the course.

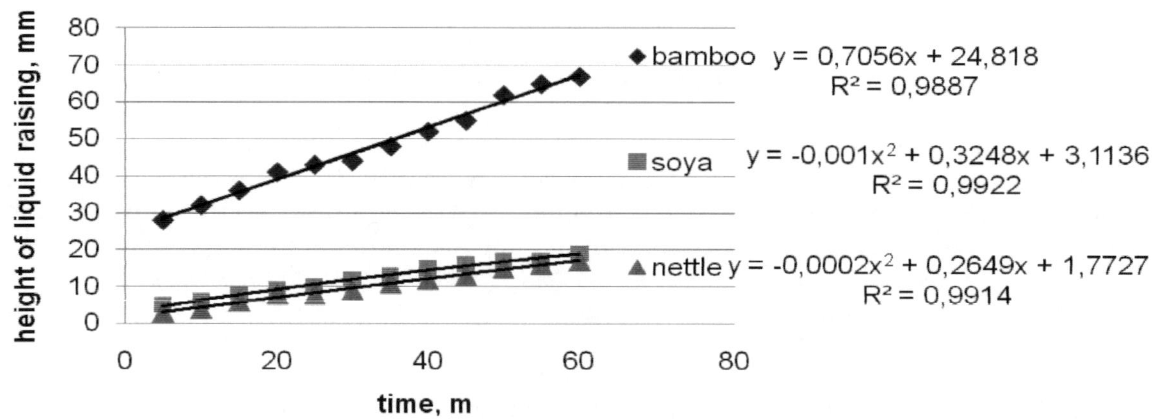

Fig. 4. Diagram of raising the liquid level of experimental models with rib structure 1+1 along the wale.

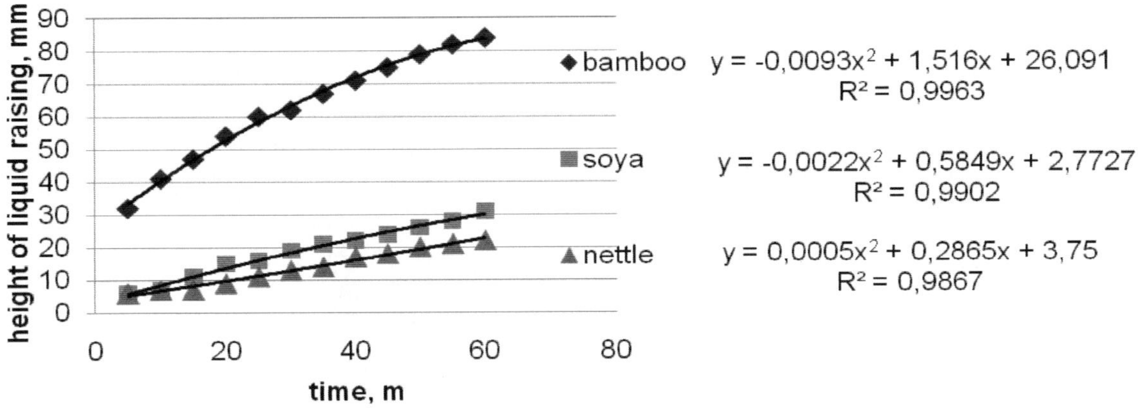

Fig. 5. Diagram of raising the liquid level of experimental models with rib structure 1+1 along the course.

Comparative analysis of influence between raw material type and liquid level shows the following. Max level is observed in the nettle models 460-530 % and along the course in models from soya yarn – 380-420 %. This tendency is observed for both fabrics with plain and rib structure.

Liquid level along the wale rises higher in experimental models with plane structure rather than with rib structure. This may be explained with peculiarities of structure formation. In the rib structure, the purl wales are located beyond the knit ones. It causes difficulties in raising liquid within capillaries.

In contrast, a very different situation is observed along the courses. In models with rib structure liquid level is higher then in models with plain structure. This is due to the larger area of surface contact between models with rib structure and the liquid.

The survey reveals that the rising liquid level depends not only on type of the raw material but also on type of structure.

Capillarity described ability of elementary sample to absorb and transfer moisture to certain height with the help of capillary forces when it is deepened into liquid for the 60 minutes [9, 10]. The survey results are presented in the histograms (Fig. 6-7).

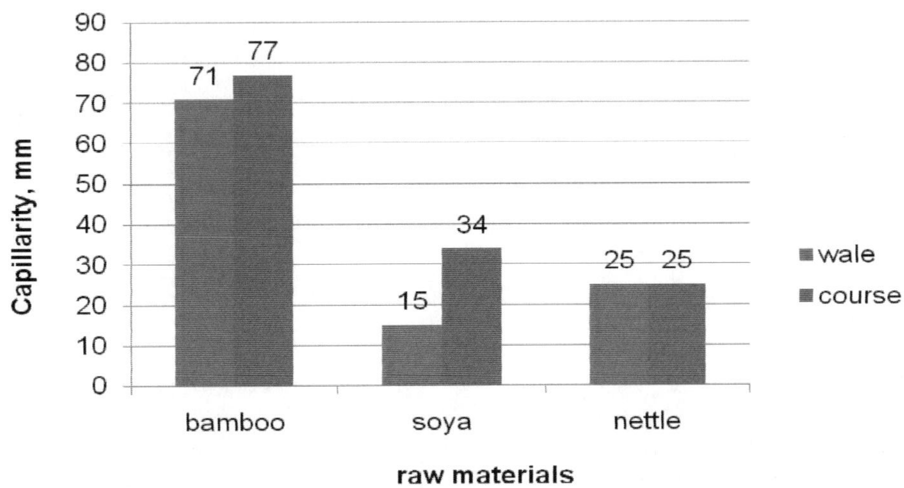

Fig. 6. Capillarity of experimental models with plain structure along the course and wale.

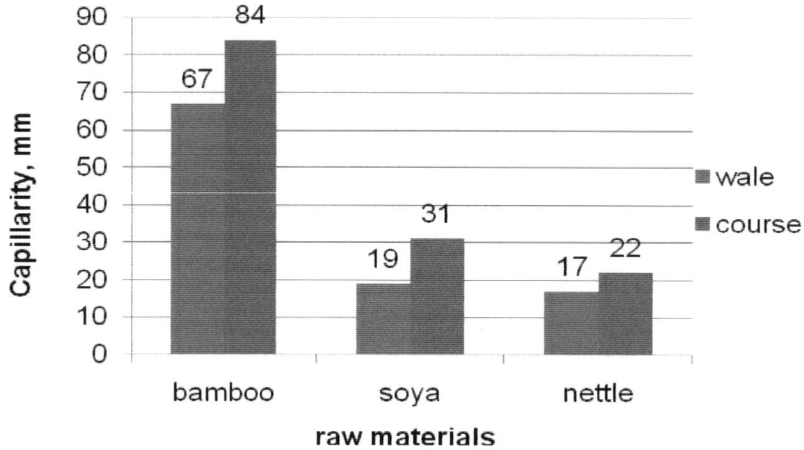

Fig. 7. Capillarity of experimental models with rib structure 1+1 along the course and wale.

As can be seen in the histograms, max capillarity along courses and along wales is observed in the experimental models from bamboo fabric.

Capillarity in experimental models with *plain structure*:

- *along wales* from soya fabric, 78.9 % less in comparison with bamboo models. In nettle fabric models in comparison with bamboo models, capillarity is 64.8 % less.
- *along courses* from soya fabric, 55.8 % less in comparison with bamboo models. In nettle fabric models in comparison with bamboo models, capillarity is 67.5 % less.

Capillarity in experimental models with *rib structure*:

- *along wales* from soya fabric, 71.6 % less in comparison with bamboo models. In nettle fabric models in comparison with bamboo models, capillarity is 74.6 % less.
- *along courses* from soya fabric, 63.1 % less in comparison with bamboo models. In nettle fabric models in comparison with bamboo models, capillarity is 73.8 % less.

Comparative analysis of capillarity of experimental models along courses and wales shows the following. In bamboo models with plain structure, capillarity along courses is 8.5 % higher than along wales. In soya models with plain structure, capillarity along courses is 127 % higher. In nettle models, capillarity is the same (Fig. 6). In models with rib structure 1+1 from bamboo fabric, capillarity along courses is 25.3 % higher, from soya fabric – 63.2 % higher, from nettle fabric – 29.4 % higher (Fig. 7). When comparing capillarity along wales in models with plain and rib structure, the following picture may be observed. Capillarity of models with rib structure along wales for bamboo fabric reduces by 5.6%, for soya fabric – increases by 26.7 %, for nettle fabric as for the bamboo one reduces by 32 % (Figs. 6-7). In experimental bamboo models with plain structure in comparison to the models with rib structure 1+1, capillarity is 9.1 % less along the courses, for models of soya fabric – 8.8 % higher, for nettle fabric – 12 % higher.

3. Results and discussion

The survey reveals that max hygroscopicity is observed in bamboo knitted models (plane structure – 23 %, rib structure – 21 %). Min hygroscopicity is observed in nettle models (plane structure – 12 %, rib structure – 11 %). Hygroscopicity depends on the structure pattern. The highest hygroscopicity is observed in knitted models with plain structure.

The survey reveals that max capillarity is observed in bamboo models (65-85 mm). Though min change in liquid level is observed in nettle models, their capillarity is only 17-25 mm.

Capillarity both along course and along wale depends on the fabric structure formation. Regardless of the type of raw material, higher capillarity is observed along the course. It is explained by direction of loop formation, so this is laterally knitted fabric. As to the type of raw material,

IOP Conf. Series: Materials Science and Engineering **141** (2016) 012013 doi:10.1088/1757-899X/141/1/012013

capillarity of models produced from nettle and soya fabrics is practically the same. Capillarity of bamboo models is 2 times higher because of bamboo fibre structure.

4. Conclusions

Knitted fabric obtained from eco-friendly raw materials has positive preventive and sometimes therapeutic effect on human body. Utilization of such materials after their life cycle is environmentally safe. However, when structure formation is generated, it is necessary to consider its influence on hygroscopicity and capillarity. Higher hygroscopicity of textile materials in items with medical and recreational effect may cause growth of pathogenic microflora under the clothing.
Information about capillarity of knitted fabric provides an opportunity to evaluate the nature of moisture spreading along the course and wale.
The study reveals the nature of change in liquid level depending on the time of experiment, knitting structure and type of raw material. With results obtained in the study, it is possible to predict hygroscopicity and capillarity of knitted fabric according to selected loop formation and type of raw material.

References

[1] Derman L 2010 Ecological design within the context of modern innovations and activities of fashion designers' in the XX and XXI centuries. [Electronic resource]. – http://derman.at.ua/publ/ ekologichnij_dizajn_v_konteksti_modnikh_innovacij_ta_ dijalnosti_dizajneriv_odjagu_khkh_xxi_stolittja/1-1-0-9
[2] Bambook. [Electronic resource]. – http://www.oven.su/netrad.htm
[3] Ecologically safer raw materials. [Electronic resource]. –http://www.pryazha.su /bambuk/
[4] This wonderful soya. [Electronic resource]. – http://knitting-pro.ru/needls-and-yarn/ yarn-for-knit/204-soy
[5] Anastasya. Handicrafts production from herbals. [Electronic resource]. –http://www. akclub.narod.ru/Krapiva/Krapiva.htm
[6] Shendikova E 2013 Exotics… in fabric. [Electronic resource]. – http://www. livemaster.ru/ topic/188453-ekzotika-v-pryazhe
[7] Nettle flax. [Electronic resource]. – http://krapivam.com/
[8] Dregulyas E 2011 Textile material science, *K.: KNUTD* 430
[9] Rybalchenko V 2010 Materials science for items of light industry. Test methods, *K.: KNUTD* 395
[10] Rybalchenko V 2006Properties of main materials in items of light industry, *K.: KNUTD*, 84

IOP Conf. Series: Materials Science and Engineering **141** (2016) 012014 doi:10.1088/1757-899X/141/1/012014

INDUSTRIE 4.0 – Automation in weft knitting technology

K Simonis[1], Y-S Gloy[1] and T Gries[1]
[1] RWTH Aachen University / Institut für Textiltechnik (ITA) Aachen, Germany

E-mail: Kristina.Simonis@ita.rwth-aachen.de

Abstract. Industry 4.0 applies to the knitting industry. Regarding the knitting process retrofitting activities are executed mostly manually by an operator on the basis on the operator´s experience. In doing so, the knitted fabric is not necessarily produced in the most efficient way regarding process speed and fabric quality aspects. The knitting division at ITA is concentrating on project activities regarding automation and Industry 4.0. ITA is working on analysing the correspondences of the knitting process parameters and their influence on the fabric quality. By using e.g. the augmented reality technology, the operator will be supported when setting up the knitting machine in case of product or pattern change – or in case of an intervention when production errors occur. Furthermore, the RFID-Technology offers great possibilities to ensure information flow between sub-processes of the fragmented textile process chain. ITA is using RFID-chips to save yarn production information and connect the information to the fabric producing machine control. In addition, ITA is currently working on integrating image processing systems into the large circular knitting machine in order to ensure online-quality measurement of the knitted fabrics. This will lead to a self-optimizing and self-learning knitting machine.

1. Introduction

The requirements for the manufacturing companies are very complex. Industry requires short response time and more efficient systems in order to remain internationally competitive. Due to the increasing number of product variants and small lot sizes the development of fast and high flexible production systems is essential, especially to small and medium-sized German enterprises. The Industry 4.0 approach can support industrial production in order to fulfil the market´s requirements [1, 2].
By technical inventions such as the Spinning Jenny and the mechanical loom, combined with the increasing availability of water and steam power, the first industrial revolution evolved facing the textile industry at the end of the 18th century. The second industrial revolution was launched at the beginning of the 20th century and was mainly caused by the mass production of goods (e.g. by Fordism). This flow production was realized by the access to electrical energy. At the beginning of the 1970s the 3rd Industrial Revolution started. This period was marked by the automation or computerization, the use of robots, new materials and of centralised control systems. The term "Industry 4.0" expresses that the industry is on the threshold of the 4th industrial revolution. Figure 1 shows the development of the industrialization [3, 4, 5].

The German location as a high-wage production region needs to stay competitive. Therefore new solutions and methods for flexible and decentralized production and control systems are to be developed. The aim of the production technology in Germany is to offer individualized products to mass production rates with a minimum planning effort. The development in the field of production

IOP Conf. Series: Materials Science and Engineering **141** (2016) 012014 doi:10.1088/1757-899X/141/1/012014

technology is extremely important for Germany to remain globally competitive. In recent years, many companies moved their production abroad in order to save manufacturing costs. The next steps of development in automation will be the integration of so-called cyber physical production systems (CPPS) using the so-called Internet of Things. The merging of real (physical) objects with information processing (virtual) objects is called cyber-physical system (CPS) .These systems will offer a cross-linking of resources, information, objects and people in order to achieve efficient and high flexible manufacturing. The cross-linking results in real-time managing and optimizing the entire production system, future companies as well as the entire logistic supply networks [6, 7].

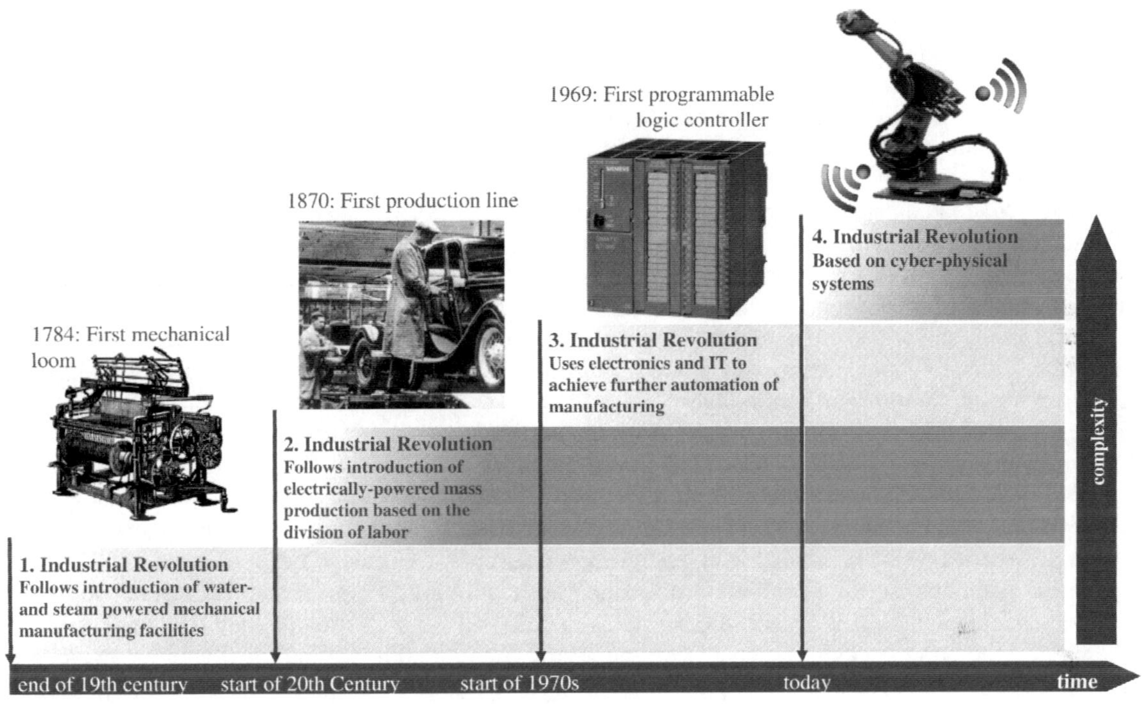

Figure 1. Development of the 4th revolution in industry [8, 9].

2. Why should the Industrie 4.0 approach be applied to the textile industry?

A survey carried out by Fraunhofer-Institut für Arbeitswirtschaft und Organisation (IAO) in 2013 show that 98,6 % of the respondents of German textile machinery manufacturers and fabric producers announced that flexibility in production will be one of the parameter of maximum significance for their future success. Flexibility means a high adjustability to unforeseen circumstances. Moreover 62 % of the respondents declare that reacting to short-term control and coordination of production orders means a high to very high effort to the companies. Especially for the apparel industry short product lifecycles, great fluctuations in demand as well as great tendencies of the customers regarding individuality require high flexibility and high efficiency in manufacturing [10, 11].

The textile production has been mainly outsourced to South-East Asia and the competitive pressure of the established German textile companies is increasing. While in 1970 there have been about 896,911 employees in the German textile industry nowadays the numbers of employees in Germany decreased to about 79,290 [12]. Challenges of industrial actors in implementing Industrie 4.0 applications are perceived regarding standardized processes, the protection of company owned knowledge as well as the skills shortage in textile [13, 14, 15]. The cross-linking of diverse individual process steps of the entire textile process chain to an overall automated process is definitely relevant for the textile industry. Because of this relevance Industrie 4.0 regarding the entire textile process chain, including the knitting process, is a big development and research topic in Germany. At the Institut für Textiltechnik der RWTH Aachen University, Aachen (ITA) the automation of the knitting process regarding Industry 4.0 applications is a big research topic. The researchers´ aim at ITA is to lower the entrance barriers for the German knitting machine manufacturers as well as the German fabric producers regarding Industry 4.0. ITA is collaborating with the Deutschen Institute für Textil- und Faserforschung Denkendorf – Zentrum für Management Research (DITF-MR), Denkendorf regarding Industry 4.0 and knitting technology. The DITF-MR for example are analysing the potential of Industry 4.0 regarding the weft knitting technology by a survey called "Strick 4.0 in Baden-Württemberg". The demand of innovation of the knitting industrial sector will be analysed in collaboration with industrial partners of the Baden-Württemberg region [16,17].

3. Project example: SpeedFactory – Automation for the sports industry

The research project SpeedFactory is publicly funded by the national program "Autonomik für die Industry 4.0". This program contributes to the implementation of the High-Tech Strategy 2020 of the Federal Republic of Germany. The program focuses on a new era of production through networking of modern information and communication technology (I & C technologies) and industrial production. At the same time versatile impetus for the creation of new innovative products, services and business models can be created using new production technologies. The project has started in October 2013 and will end after a three - year period in 2016. Each partner will provide its individual know-how in their respective fields of work to the project plan. The consortium consists of the following partners, who contribute both in industry and research: Johnson Controls GmbH, Burscheid - a leading automotive supplier. The project team of the adidas Group, Herzogenaurach consisting of experts from the adidas Innovation, Global IT innovation and sourcing. Pfaff KSL Keilmann Sondermaschinen GmbH, Lorsch – Experts in manufacturing of individual robot production plants belong to the consortium as well. In addition, the fortiss GmbH, Munich is part of the consortium.

The aim of this project is to encourage the development of autonomous systems. Germany needs to expand its leading position as an industrial centre for new and pioneering internet-based technologies. The research covers topics such as production logistics, basic cognitive technologies, human-computer interaction (HCI) and 3D in industrial applications. The project's objective will be reached by creating solutions for the automatic production and handling of limp fabrics. Therefore robot and handling systems will be deployed and developed. Within the research project, the ITA is focusing on the entire textile process chain, especially on the textile fabric manufacturing, the handling as well as on the joining process [18, 19].

4. Cross-linking of the knitting process with upstream process steps – Smart bobbin [13]

By the use of these tracking and tracing systems fabric producers as well as consumers are able to follow the products history. Therefore, the following data are to be transferred during the production process [20]:

- Where is the product stored, how much material is left?
- Which production steps have been worked out already?
- In which condition is the product at the moment?

In order to save and transfer information the use of optical codes with an estimated ratio of 75 % is widely spread. Because of decreasing prices, the small size and the increasing performance the ratio of applications for the Radio-Frequency Identification (RFID) Technology is increasing as well [21-24].

The RFID Technology offers high potential regarding Industry 4.0 applications. By the use of RFID process and/or product data can be transferred from e. g. machine to machine. Research at ITA show that yarn quality during the spinning process can be detected using spinning machine embedded sensors. The information of the location of errors along the yarn length is saved on a RFID chip which is placed into the corresponding bobbin. By the use of the RFID chip and the corresponding receiving device these error data will be transferred to the fabric producing machine – knitting or weaving machines. Investigations at ITA show that as soon as the smart bobbin (weft bobbin) is located at the weaving machine the data can be transferred to the machine's Soft PLC in order to control online the production speed depending on the yarn quality characteristics (errors). The development of this concept has been honored by the Wilhelm-Lorch-Stiftung in 2014 [25, 26].

ITA is currently working on transferring the concept to the large circular knitting technology in order to recognize the characteristics of each bobbin characteristics in the bobbin creel (yarn material and fineness, color and yarn length, yarn tension) (see Figure 2). The online recognition of the diverse yarn characteristics in the creel related to the knitting system positioning will be especially relevant for the jacquard knitting technology. ITA is expecting to reduce the setting time after changing yarn material in the bobbin creel by about 20 %. In case of knitted product changes the operator is able to react more flexible to the pattern and/or yarn changes because the operator is no longer forced to collect the information manually. In times of the demographic change and a rising lack of educated operators systems of data transformation help to increase the international competitiveness of the German knitting industry.

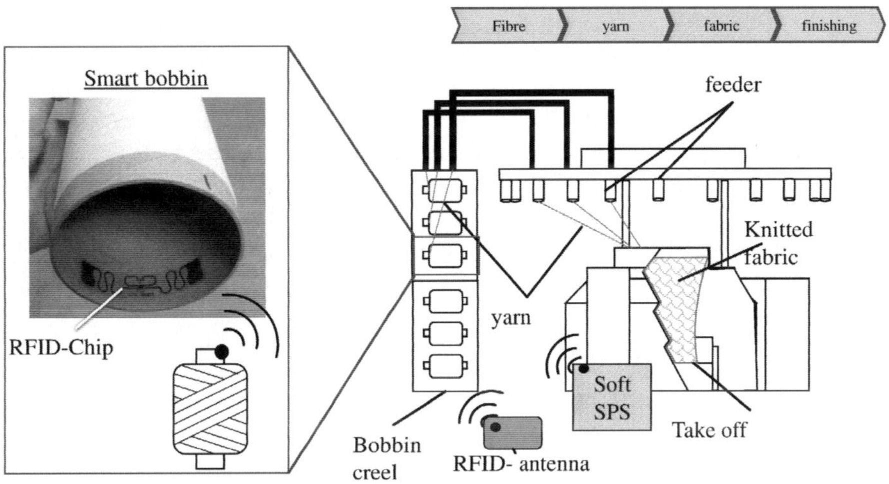

Figure 2. Smart bobbin – concept of RFID-technology used in large circular knitting technology [27].

Besides the smart bobbin research at ITA, there have been investigations in physically knitting QR-Codes in order to transfer fabrics' information. By this technology the knitted fabric can be provided with product, downstream or upstream process information. In this case no additional chip or code is to be purchased including the risk of getting lost or being damaged in downstream process steps. The readability and sizes of the QR-Code directly corresponds on the one hand with the amount of the machines' needles (machine gauge) and the yarn material and fineness. On the other hand the interaction of the knitting process parameters speed, take-off tension and sinker height influences the codes' dimension and therefore the readability of the code.

5. Assistance systems - Augmented Reality (AR) in knitting

ITA is also working in including the operator into the field of Industry 4.0. The development of human-machine interfaces (HMI) such as Augmented Reality (AR) based assistance systems supports the operator. The setting of the process parameters according to the used material and desired product specification are worked out by operators according to the gained experience and knowledge. Especially in times of demographic change and staff shortage for machine operators in Germany it is increasingly important to systematize this individual process knowledge in order to support inexperienced employees. These assistance systems can be used on the one hand for the logistic purpose and in production, and on the other hand as assistance in maintenance operations [28].

IOP Conf. Series: Materials Science and Engineering **141** (2016) 012014 doi:10.1088/1757-899X/141/1/012014

6. Image processing in knitting

Optical characteristics, such as the loop uniformity and surface condition, are crucial quality criteria of knitted fabrics. The human eye can detect deviations of less than 20 µm. In the field of technical textiles (e.g. flame resistant underwear, fencing suits, skiing underwear) and fine knit-wear (e.g. socks) single, selective defects in the fabric can lead to fabric rejections. A fabric defect is any abnormality in the fabric that hinders its acceptability by the consumer. Defects are caused either by inhomogeneous yarn material mixture, fineness and tension or by false setting of knitting process parameters such as speed [rpm], sinker height [mm], yarn tension [cN], take-off tension [N] or the width of the spreader [mm].

In particular, undetected defects with spatial extent in the production direction (e.g. color defects, pattern error, laddering) are critical because of the impact on in some cases the entire production batch. Moreover, repeating errors in the fabric texture, such as striper defects or loop distortion are hardly to be differentiated from an intended pattern texture. These defects just can be detected during production by trained and experienced operators. Such visual inspection tasks are perceived as monotonous and tiring to the operators [29]. Moreover, the operators in large circular knitting manufacturing companies supervise several machines at the same time, so the operators cannot guarantee a 100 % quality observation. Therefore, the manually inspection should be substituted by image processing systems. The digital image processing allows outsourcing the visual inspection by computer technology. In this way, complex and repetitive tasks of error detection, such as counting, measuring and defect detection, are solved by computer systems. Although in other technologies, like the steel or automotive industry, visual detecting systems are common because of the practical advantages, there has been minimal application of image-processing systems in the knitting industry, yet.

ITA is collaborating with the Institute of Imaging and Computer Vision of the RWTH Aachen University investigating the possibilities in automated detection of defects in knitted fabrics. Striper defects for example appear in case the tension of at least one yarn along the knitting cylinder differs from the others. Small differences in yarn tensions are hardly to be detected by the human eye during production, but do affect the fabric quality. The fabric´s treatment during the finishing process – especially during the dyeing process – leads to an increasing visibility of the striper defects. As a result, the fabrics are to be declared as substandard goods or rejects (Figure 3).

Nowadays, some defects, such as stripers resulting from little differences in yarn tension, are difficult to detect visually and during production even by knitting experts. As a result of the collaboration with the Institute of Imaging and Computer Vision of the RWTH Aachen University, an image processing method has been developed allowing to significantly enhancing stripers in knitted fabrics. Images are first pre-filtered to increase contrast and homogeneity. Subsequently, spectral filters and correlation applications are applied to enhance visibility of the stripers. The resulting images can be displayed on-screen to improve the reliability of manual inspection or processed further in order to allow training of an automated, machine learning-based classifier. Once trained, the classifier is able to detect stripers in new images with high reliability and without any human interaction.

In the current downstream fabric inspection the open-width fabric is examined optically. Error avoidance during the knitting process is not possible with the current downstream fabric inspection. Fabric areas with failures are recognized and rejected after the downstream fabric inspection. The retrofit operation in knitting plants includes not only the yarn change (approx. 30 minutes), the pattern change (0.5 days with mechanical equipment , for 1 hour at jacquards) in addition to patterning and material-dependent adjustment process of the knitting parameters (up to 1 day).

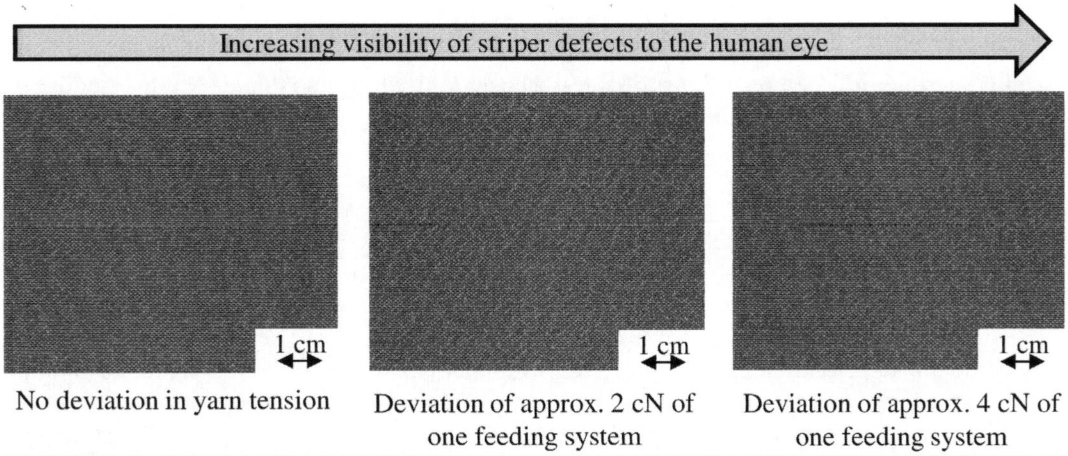

Machine:
Large circular knitting machine, BSM 3.0, Beck GmbH, Albstadt (E24, 30"), 18 systems activated, sinker 0.5, Speed 30 rpm
Material:
Polyester yarn, 68 filaments, 293 dtex, texturised, high intermingled

Figure 3. Increasing in striper defect visibility by deviated yarn tension of one of 18 feeding systems

In addition to defect detection, image processing systems can also support the prototype development of new knitted products. Although program software tools such as the M1 software of the company H. Stoll AG & Co. KG, Reutlingen is able to visualize digitally the different pattern modules, the calculation of the resulting dimension of two or more pattern combinations is not possible in industry yet. Therefore, the prototype development of knitted products is worked out by trial and error. This development process can take up to a year depending on the pattern complexity of the product. Investigations at ITA show that the mesh density of a pattern is depending on the surrounding pattern (Figure 4). This influence can either be simulated ex ante or visually measured online using a camera system, image processing and a self-learning algorithm. The knitted resulting dimensions of the single pattern regions are measured online and compared to the target dimension. The self-learning system will enable to change the knitting program in terms of adding or reducing the amount of loops per pattern region. Finally the cross-linking to the machine control will ensure the production adjusted knitting program [30]. A target-actual comparison provides the measurable deviation from the desired structure so that the necessary adjustments can be performed.

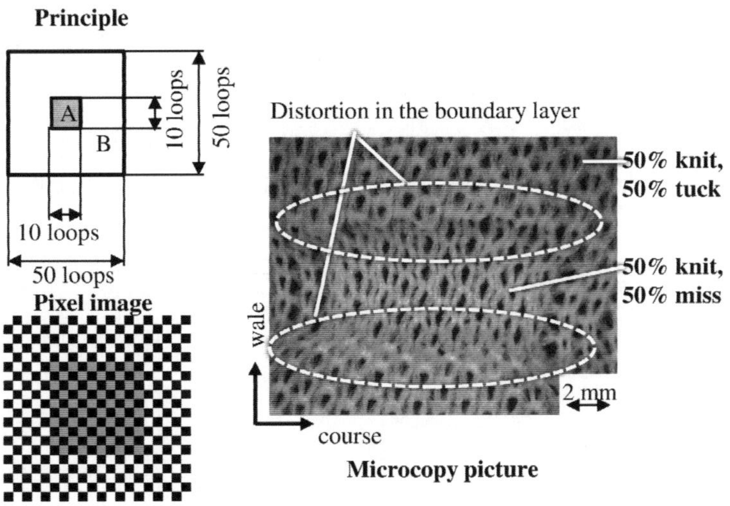

Figure 4. Exemplary distortion of the inner pattern region by the surrounding of another pattern region

7. Online and reproducible control of the spreader width in large circular knitting machines

In the large circular knitting technology the knitted tube is formed by a so-called spreader to a double-surfaced knitted fabric. As a result, this double-surfaced knitted fabric is guided to the middle of two take-off rolls and is either wound to a cloth bale or folded into a cloth box. The adjustment of the dimensions of the spreader affects significantly the properties of the knitted fabric in terms of loop size and loop distortion. Additionally investigations at ITA show that the take-off tension among other process parameters significantly influences the mesh density in wale direction (Figure 5). Currently, the spreader width is manually adjusted according to the knitting patterns and to the yarn elasticity. This adjustment is based on the experience and knowledge of the operator in charge. Moreover, due to the spreader design, there is no possibility to adjust the spreader width in a reproducible way. Two bars are literally to be pulled apart from each other by the operator. Therefore, the width optimum according to the selected material and pattern is not known. Furthermore, even if the optimum is reached by accident, the reproducibility of the corresponding setting parameters is not given. The concept at ITA is to measure the fabrics pressure to the spreader bars by pressure sensors. The pressure of the fabric directly corresponds with the fabric characteristics regarding elasticity, tensile strength and mesh density. Figure 6 shows the results of a drape-test of five different knitting patterns in collaboration with the Faserinstitut Bremen e. V., Bremen, Germany. The diagram shows that the fabric resistance is depending on the combination of the three existing pattern or cam elements in large circular knitting: miss, tuck and miss.

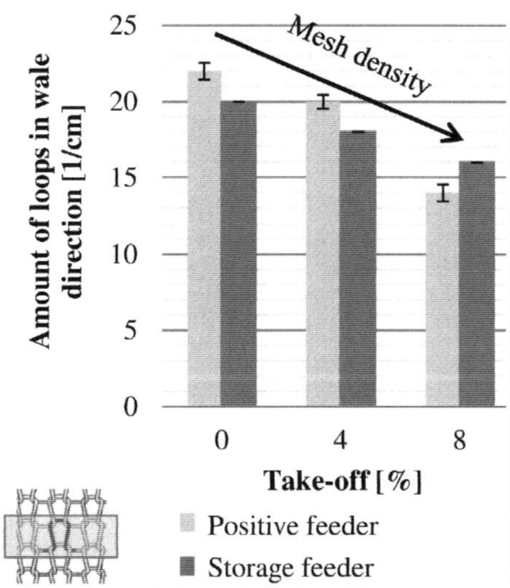

Take-off setting [%]	Microscopic pictures	Mesh density [1 / cm²]
0	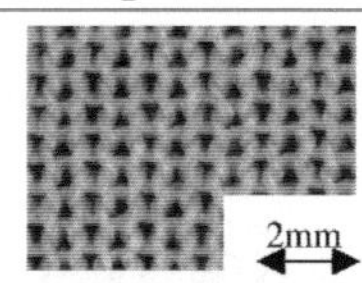	148
4	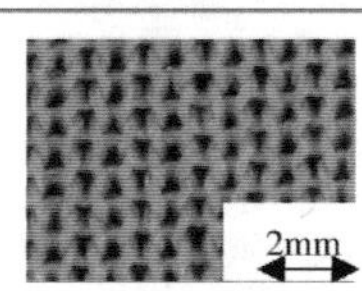	146
8	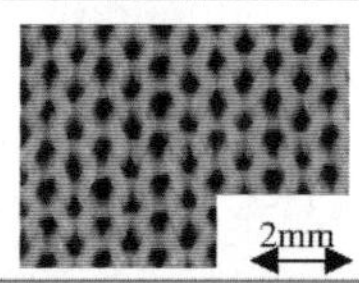	124

Machine:
Large circular knitting machine, BSJM 1.3, Beck GmbH, Albstadt (E16, 19"), 4 systems activated, sinker 0.3, Speed 28 rpm
Material:
Polyester yarn, 68 filaments, 250 dtex, texturised, high intermingled, Single Jersey

Figure 5. Influence of the take-off settings on the mesh density in single jersey fabrics.

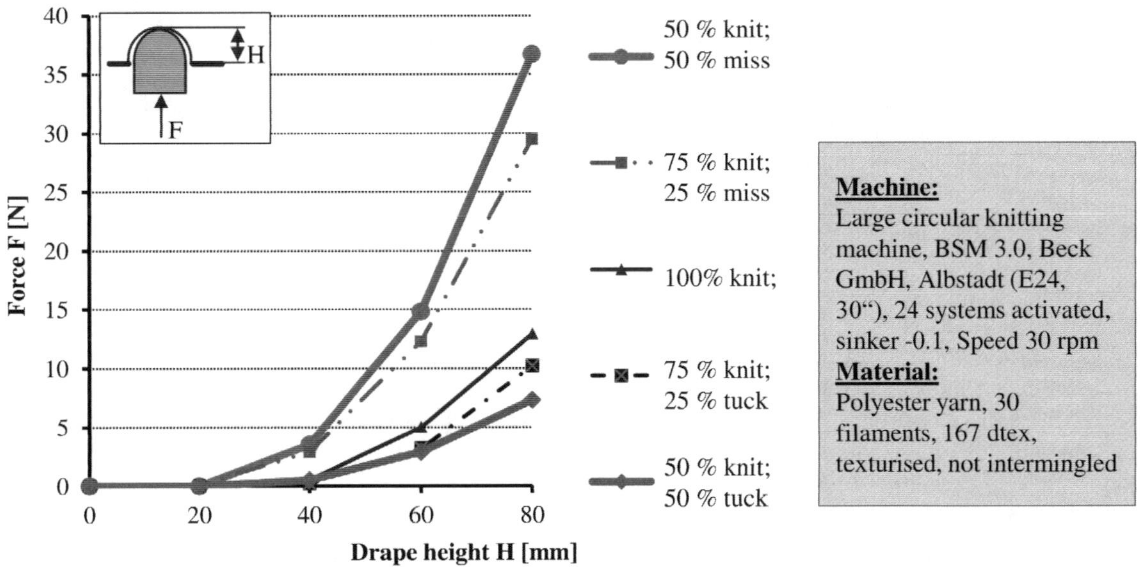

Figure 6. Drape-test results of different combination of cam elements – in collaboration with Faserinstitut Bremen e. V., Bremen.

In addition to the varying resistance the circumference length of the circular knitted tube also differ by the chosen pattern. The combination of cam elements results in varying mesh densities in wale and course direction. According to that the porosity of patterns also varies the resulting pressure of the fabric to the spreader. That is the reason why the spreader of the circular knitting machine is to be adjusted according to the desired product. The reproducibility of fabrics' quality by adjusting the spreader settings gets more complicated when lots of pattern changes are introduced by using the electronic jacquard technology. The variation in pressure of the fabric can be controlled by the new spreader control concept. As a result ITA ensures the online and product specific, reproducible settings of the spreader width, even for jacquard knits. In order to ensure the acceptance by fabric and knitting machine manufacturers the price is aspired being under 1000 € per system (development status).

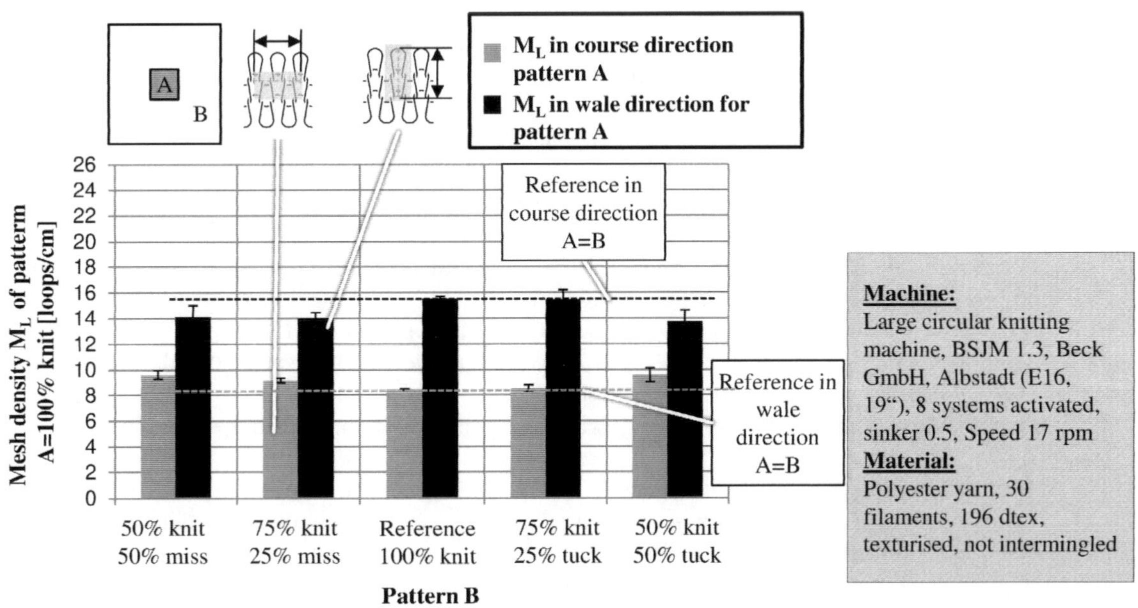

Figure 7: Variation in mesh density according to selected combination of two patterns

8. Summary

The Industry 4.0 and automation approach is relevant to ensure the international competitiveness of the German knitting industry because of the demographic change and experience-based knowledge in production. Currently, the correspondence data of the knitting process parameters are rarely systemized in industry. The collecting and systemization of process data is a basis requirement for automation and intelligent production. Aim of this intelligent production using self-learning and self-optimizing systems in knitting is to increase product quality, decrease fabric rejection costs and support the operator during manufacturing.

9. References

[1] http://www.plattform-i40.de/I40/Navigation/DE/Industrie40/ChancenIndustrie40/chancen-durch-industrie-40.html, 06.05.2016

[2] Promotorengruppe Kommunikation 2013 *Umsetzungsempfehlung für das Zukunftsprojekt I40* (Frankfurt am Main).

[3] Sendler U 2013 *Industrie 4.0* (Springer Vieweg).

[4] Gloy Y-S 2014 *Industrie 4.0: the future of the textile production?* vol 20 (Melliand International)

[5] Kagermann H, Lukas W 2011 *Mit dem Internet der Dinge auf dem Weg zur 4. Industriellen Revolution* vol 13 (VDI Nachrichten)

[6] Brecher C 2011 *Integrative Produktionstechnik für Hochlohnländer* (Springer-Verlag Berlin)

[7] http://www.plattform-i40.de/I40/Navigation/DE/Industrie40/ChancenIndustrie40/chancen-durch-industrie-40.html, 06.05.2016

[8] Bauer W, Schlund S, Marrenbach D; Ganschar O 2014: *Industrie 4.0 - Volkswirtschaftliches Potenzial für Deutschland, Studie des Bundesverband Informationswirtschaft*, Telekommunikation und neue Medien e. V. (BITKOM), Berlin

[9] Promotorengruppe Kommunikation 2013: *Umsetzungsempfehlung für das Zukunftsprojekt I40*

[10] Spath D, Ganschar O, Gerlach S, Hämmerle M, Krause T, Schlund S 2013 *Produktionsarbeit der Zukunft - Industrie 4.0* (Fraunhofer-Verlag)

[11] Chancen und Herausforderungen durch Industrie 4.0; http://www.bayern-innovativ.de/textil2015/presse/pressemitteilung? 2015

[12] Jacoangeli M 2016 *Konjunktur 03.2016* (Textil und Mode, Gesamtverband der deutschen Textil- und Modeindustrie e.V., Berlin)

[13] Spath D, Ganschar O, Gerlach S, Hämmerle M, Krause T, Schlund S 2013 *Produktionsarbeit der Zukunft - Industrie 4.0* (Fraunhofer-Verlag)

[14] Gloy Y.-S, Greb C, Gries T 2013 *Industrie 4.0: a (r)evolution for the textile industry?* (Hillmer, Janine (Ed.): Proceedings of the 7th Aachen-Dresden International Textile Conference, Aachen November 28-29)

[15] Promotorengruppe Kommunikation 2013 *Umsetzungsempfehlung für das Zukunftsprojekt I40*

[16] https://mfw.baden-wuerttemberg.de/de/service/presse-und-oeffentlichkeitsarbeit/pressemitteilung/pid/ministerium-fuer-finanzen-und-wirtschaft-foerdert-studie-strick-40-in-baden-wuerttemberg/

[17] http://www.gesamtmasche.de/news/nachricht?n=650&cat=&seite=1

[18] Simonis K, Lutz V, Gloy Y.-S, Gries T 2016 *SpeedFactory : Autonomik für die Sportindustrie*, (WTI) WTI-Dokumentnummer 20160200338

[19] http://www.avronline.de/ueber+speedfactory+und+neue+wertschoepfung+.154243.htm, 03.05.2016,

[20] Bartneck N 2008 *Prozesse optimieren mit RFID und Auto-ID* (Publics Corporate Publ.)

[21] Bartneck N 2008 *Prozesse optimieren mit RFID und Auto-ID* (Publics Corporate Publ.)

[22] Kern C 2006 *Anwendungen von RFID-Systemen* (Springer Verlag)

[23] Kern C 2006 *Anwendungen von RFID-Systemen* (Springer Verlag)

[24] Özel H 2008 *Bedeutung und Anwendungsgebiete des RFID-Systems* (Diplomica Verlag)

[25] Winkel B 2014 *Entwicklung und Bewertung eines Radio Frequency Identification – Informationsinfrastrukturkonzepts entlang der textilen Kette für die SMART Factory* (Bachelor Thesis at Institut für Textiltechnik der RWTH Aachen University, Aachen)
[26] http://www.wilhelm-lorch-stiftung.de/foerder/pages/2015/bernd_winkel.html, 06.05.21016
[27] Simonis K, Gloy Y.-S 2015 *Strick 4.0 : Revolution der Maschenindustrie* vol 3 (Masche)
[28] http://www.ita.rwth-aachen.de/3-f-und-d/kurzberichte/2016/2016_02_02%20SozioTex_Saggiomo_Projektsteckbrief.pdf, 06.05.2016
[29] Langenbach R 2002 *Methoden der Farbbildverarbeitung für die automatische Oberflächeninspektion* Universität-GH Siegen, Dissertation
[30] Simonis K Gloy Y.-S Gries T 2016 *Selbstlernende Systeme für die Strickerei von Morgen* (Melliand Newsletter Januar 2016) URL:http://textination.de/de/document/1145304385384394/2.0/ITA-%20selbstlernende%20Systeme%20Strickerei%20von%20Morgen.pdf

Acknowledgments

Supported by:

Federal Ministry for Economic Affairs and Energy

on the basis of a decision by the German Bundestag

The Institut für Textiltechnik of the RWTH Aachen University, Aachen likes to thank the Federal Ministry for Economic Affairs and Energy for the supporting the project “SpeedFactory – Autonomik for the Sports Industry”.

Recent developments in multi-layer flat knitting technology for waste free production of complex shaped 3D-reinforcing structures for composites

W Trümper, H Lin, T Callin, Q Bollengier, C Cherif and S Krzywinski

Institute of Textile Machinery and High Performance Material Technology at Technische Universität Dresden, 01062 Dresden, Germany

E-mail: wolfgang.truemper@tu-dresden.de

Abstract. Constantly increasing prices for raw materials and energy as well as the current discourse on the reduction of CO2-emissions places a special emphasis on the advantages of lightweight constructions and its resource conserving production methods. Fibre-reinforced composites are already seeing a number of applications in automobile, energy and mechanical engineering. Future applications within the named areas require greater material and energy efficiency and therefore manufacturing methods for textile preforms and lightweight constructions enabling an optimal arrangement of the reinforcing fibres while in the same time limiting waste to a minimum. One manufacturing method for textile reinforced preforms fulfilling quite many of the named requirements is the multilayer weft knitting technology. Multilayer weft knitted fabrics containing straight reinforcing yarns at least in two directions. The arrangement of these yarns is fixed by the loop yarn. Used yarn material in each knitting row is adaptable e. g. according to the load requirements or for the local integration of sensors. Draping properties of these fabrics can be varied within a great range and through this enabling draping of very complex shaped 3D-preforms without wrinkles from just one uncut fabric. The latest developments at ITM are concentrating on the development of a full production chain considering the 3D-CAD geometry, the load analysis, the generation of machine control programs as well as the development of technology and machines to enable the manufacturing of innovative net shape 3D-multilayer weft knitted fabrics such as complex shaped spacer fabrics and tubular fabrics with biaxial reinforcement.

1. Introduction

The direct manufacturing of geometry and load adapted textile semi-finished products or preforms for setting up of complex fibre reinforced plastic composites (FRP) is limited using currently available technical textile solutions. However, the need for such preforms increases with the growing range of applications of FRP. In addition to aerospace, automotive or applications in wind energy there is an increasing demand for the manufacture of sports and leisure goods, medical devices or components in mechanical and plant engineering. These parts are currently manufactured from plain textile structures during extensive manual forming and draping processes. Thus the manufacturing of FRP is nowadays very expensive at a very limited level of reproducibility.

One main goal of various research projects at the Institute of Textile Machinery and High Performance Material Technology (ITM) is therefore the development of textile technologies, such as flat knitting technology, for a direct manufacturing of 2D and/or 3D net shape preforms. For a wide industrial implementation of the results with best possible consideration of component requirements a digital connection between the 3D CAD part geometry and the knitting pattern is necessary. As a result manufacturing requirements, stress-related reinforcement yarn arrangement and part geometry can be considered in a very early phase of future FRP component development. In addition to the development of the necessary manufacturing technology and the multi-layer knitted fabrics (MLG) itself, the focus is thus the process chain development from CAD model to knitting patterns

Below the possibilities are presented how innovative net shape textile reinforcing structures can be manufactured by means of flat knitting technology. Today pure knits are manufactured as a complete article in one piece [1 – 3]. The disadvantages for the use of such structures in FRP are the structurally high elongation and the low strength. With the developments of the flat knitting technology and especially the MLG at ITM these disadvantages can be overcome.

2. Multi-layer weft knitted fabrics (MLG).

2.1. Set-up of MLG

MLG are based on a single or double-jersey binding with integrated straight reinforcing fibres in course (weft yarn) and wale direction (warp yarn) [4, 5]. Figure 1a shows an MLG with one warp and one weft yarn layer. The number of layers can vary depending on the requirements. From machine point of view currently the number of layers is limited to a maximum of 11 reinforcing layers (Figure 1 c) [6].

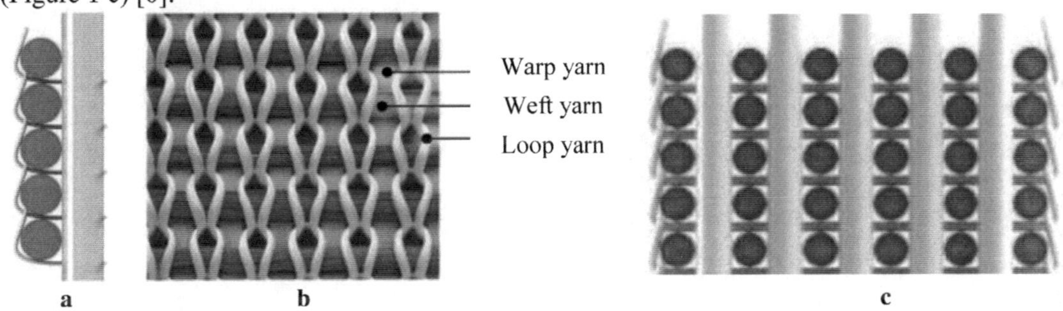

Figure 1 . Set up of multilayer weft knitted fabrics a) cross section 2 layer fabric, b) front view, c) cross section 11 layer fabric [6].

The weft yarns are integrated by means of yarn guides. For the warp yarns special warp yarn bars were developed to enable the simultaneously feeding of a high number of warp yarns towards the stitch-forming area. The loop yarns are used for the fixation of the straight reinforcing yarns. The biaxial arrangement of the straight reinforcing yarns within the MLG allows the extensive utilization of their mechanical properties. MLG also exhibit a structural dependent excellent draping behaviour which is also locally adjustable in a wide range by means of the loop length [7].

2.2. Economic analyses

Before the further developments of the 2D and 3D MLG technique are presented, first economically benefits achievable by the use of MLG structures during the manufacturing of FRP components may be revealed. A closer look on the current cost of FRP components shows that about 25 % are accounted for the textile semi-finished products and another 50 % for the mostly manually done sequential preforming [8-11]. This means up to 75 % of the component costs are directly related to the textile reinforcing structures.

Part of the cost of the textile structures are e. g. machine costs but also materials costs. According to different companies and depending on the component geometry currently about 40 to 50 % of the material is being scrapped [8].

A first economic analysis (Table 1) with the conservative assumption that through the use of net shape textile reinforcing structures a lowering of the use of materials by 40 % and the effort for the manual preforming by 50 % can be achieved reveals a reduction of the cost of preform for CFRP components by approximately 36 %.

Table 1: Comparison – cost of preform manufacturing

	Textile structure			Preforming	Preform	Change
	CF-Material		manufacturing			
	[kg]	[€]	[€]	[€]	[€]	[%]
Sequential preforming	1,8	36	9	90	135	
Net shape structures	1,3	26	15	45	86	-36

Assumptions for the calculations are a mass of the CFRP component of 2 kg at 55 % fibre volume content. The cost of the preform manufactured by sequential preforming were set to 135 € based on references [10, 11]. The numbers show the enormous potential for reducing manufacturing costs of FRP components by the use of material-efficient net shape semi-finished textile reinforcing structures. The knitting technique, especially the MLG technique is currently one of the few textile techniques enabling a direct manufacturing of net shape 2D and/or 3D reinforcing structures for high-performance FRP applications. Following the different options to simplify the preforming process by the use of MLG are presented in detail. In addition to the mentioned direct manufacturing of net shape structures this also includes the integration of a high number of reinforcing layers into the MLG.

2.3. Material and cost efficient preforming by use of MLG

2.3.1. High number of reinforcing layer

The effort for the production of textile preforms for FRP components essentially depends on following factors:

- Number of individual layers of the preform and
- Expenses for cutting, lay-up and forming of the final geometry.

The number of textile structures necessary to set up a preform can be significantly minimized by the use of semi-finished products with a high number of reinforcing layers. In Figure 2 an example is shown, indicating the number of necessary textile structures to set up a preform consisting of 12 reinforcing layers depending on the number of layers per structure.

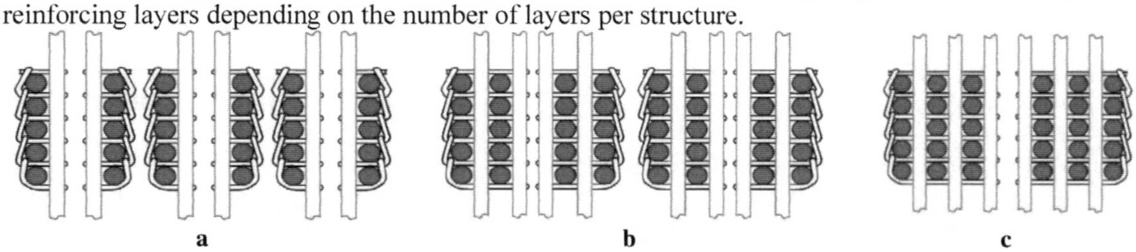

Figure 2. 12-layer preform set up using a) 2-, b) 4- und c) 6-layered multilayer weft knitted fabrics

Studies at ITM show that the number of reinforcing layers per each MLG has only a small influence on the performance of the manufacturing machines. This means that semi-finished products in accordance with Figure 2 c can be produced at a level of about 1/3 of the production time compared to the semi-finished products according to Figure 2a. Additionally only two semi-finished products need to be cut to shape and draped during the set-up of the preform. The draping and thus the formability of the textile structures are still set depending on the loop length. A general assessment of the drapability can be made on the base of the results of comparative studies on the shear properties (Figure 3).

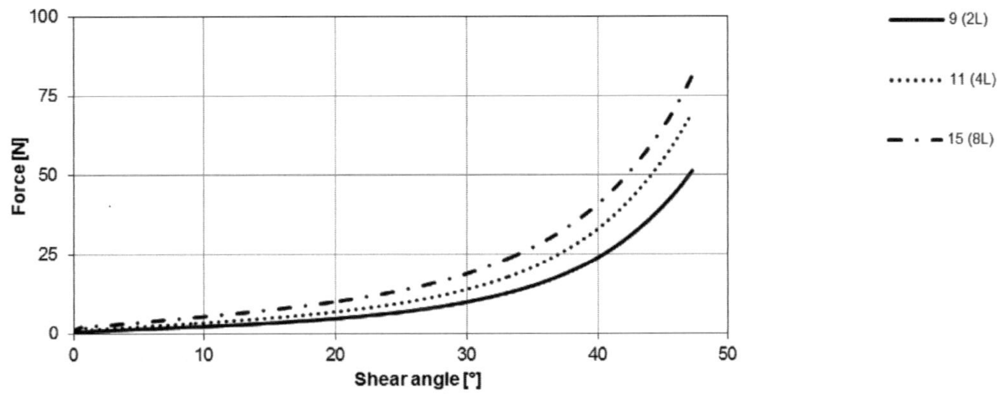

Figure 3. Test results of shear tests at multilayer weft knitted fabrics with different number of reinforcing layers

From the diagram in Figure 3 it can be seen that the forces required for the achievement of a certain shear angle increases with number of layers in the semi-finished product. Due to the fact, that this is a slight increase a negative influence on the forming of the textile structures during the set-up of the preform is not expected.

Essential for the use of textile structures for the FRP production is next to the manufacturing expense the mechanical performance of the composites made using them. To evaluate the mechanical properties extensive tests on thermoplastic composites made of semi-finished products with different number of reinforcing layer were (Figure 3) performed. All warp and weft yarns used in the biaxial - reinforced MLG are hybrid yarns from glass fibre (53 vol.- %) and polypropylene (47 vol.-%) with a total fineness of 840tex. The warp and weft density is equal. The spacing of the warp to warp and weft of about weft is approximately 3.6 mm. As loop yarn plied glass and polypropylene filaments are used. The diagram in Figure 4 shows the tensile strength and the elastic moduli of the thermoplastic composites measured in tensile tests.

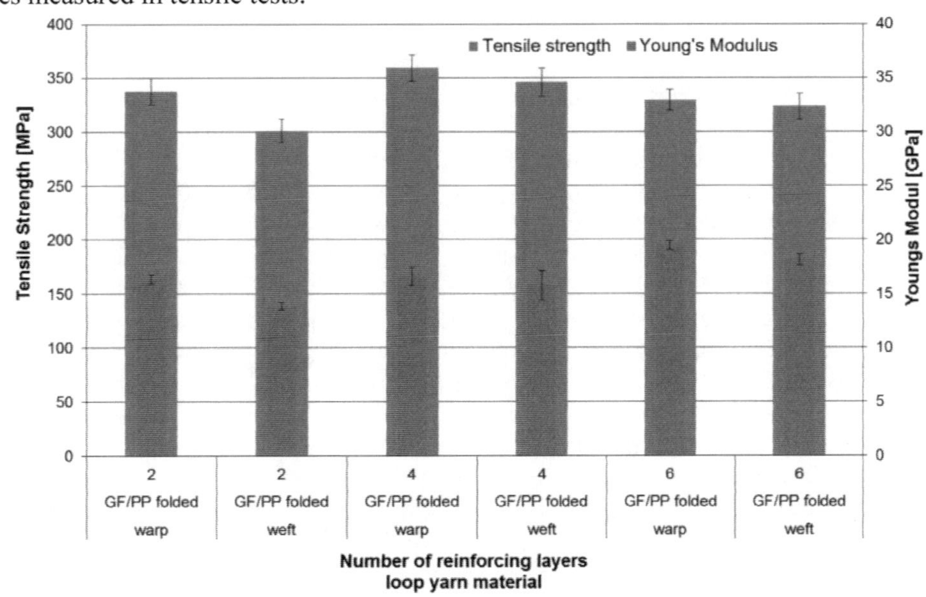

Figure 4. Tensile test results at composites specimens made from multilayer weft knitted fabrics with different number of reinforcing layers per fabric

From Figure 4 it can be seen, that the change of the number of reinforcing layers within the MLG, has only a little influence on the mechanical properties of the composites. The overall conclusion is a slightly rise of the Young's modulus while the tensile strength exhibits nearly the same values. This trend is very likely due to the lower loop yarn proportion in the composites produced from MLG with a high number of reinforcing layers. The slightly higher mechanical properties in the warp direction are caused by the manufacturing inherent higher warp yarn tension during the production of MLG. The results confirm that MLG with low and high number of reinforcing layers exhibit the same mechanical performance. Thus it is possible to reduce FRP cost by using MLG with a high number of reinforcing layers while keeping the mechanical performance.

2.3.2. 2D- and/or 3D-net shape preforms

While setting up preforms from textile structures with a high number of reinforcing layers less single textile fabrics have to be used and hence the preforming effort during production of FRP components can be significantly reduced (compare chapter 2.3.1.). However, the textile structures still have to be extensively and usually manually draped according to the final 3D-geometry. In addition, there arises a FRP-geometry dependent waste of up to 40% of the material used. Therefore, a key objective in the field of textile technology is the development and implementation of methods for a direct preforming of the 2D-outer contour and/or the 3D-geometry during textile production.

While using weft knitting technology for the direct preforming shaping possibilities such as widening, narrowing or flechage also known as short row knitting are preferable used (Figure 5).

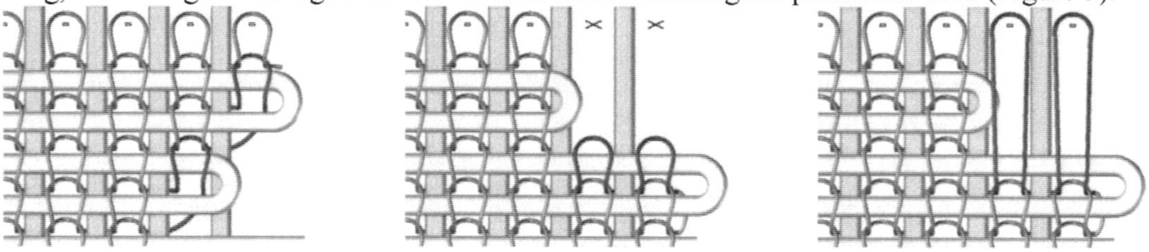

Figure 5. Net-shape knitting technology a) widening, b) narrowing, c) flechage [12]

Narrowing and widening are used for the realization of the outer contour of knitted fabrics, whereas flechage is used for the creation of the 3D geometry [5]. Anyway all these shaping possibilities are used for the adaptation of the loop structure.

An additional important requirement for the manufacturing of biaxial reinforced net shape MLG is the feeding of the required yarn lengths of warp and weft yarns. The feeding of the yarns must be carried out with a high accuracy according to the pattern and corresponding to the component geometry. Another important requirement to achieve a high geometric accuracy is the possibility to feed every warp yarn at each needle with an individual adjusted length. At the same time the therefore necessary systems must ensure the transport of the yarns from the yarn storage to the stitch forming area, the adjustment of the fabric tension required for the knitting process and the withdrawal of the textile structure produced of the stitch-forming area. Figure 6 shows an according to these specifications developed 3D take down system with an additional active warp yarn feeding device which is arranged in before the stitch-forming area.

a

b

Figure 6. Net-shape knitting a) machine b) detail warp yarn feeding unit and segmented 3D take down system [6].

The 3D – take down system is equipped with 30 single motor driven take down segments which enable individual take down of a very small fabric region [13]. In addition to the implementation of the take down system into the space available at flat knitting machines another major challenge is to provide a flexible control of the take down system. This is necessary to enable the manufacturing of different contours or geometries. The necessary constructive work and technological studies were successfully completed.

Figure 7 shows the also necessary and parallel within the SFB639 developed continuous process chain that starts with the 3D-CAD geometry which is converted into a 2D outline to generate following the drive and control programs for the knitting machine. Also shown are exemplary and by using the newly developed machine technology produced net shape MLG.

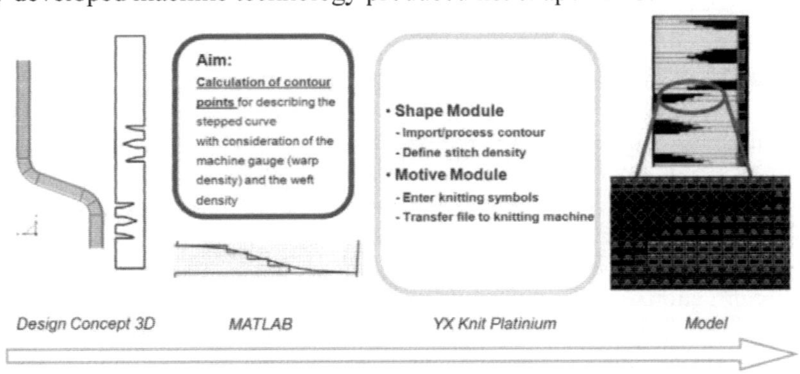

a

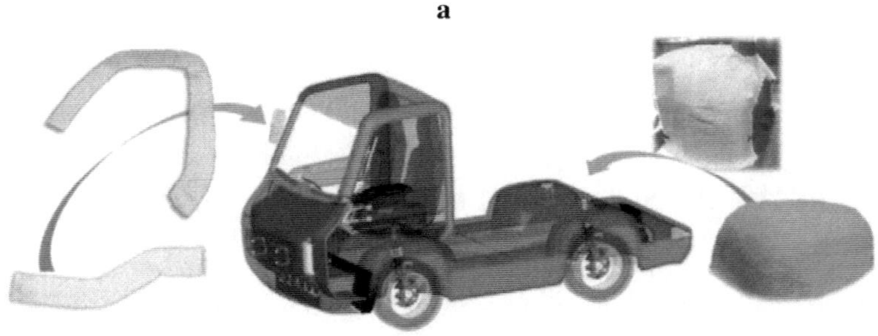

b

Figure 7. Weft knitted net-shape preforms a) production chain [14], b) demonstrator [SFB/CRC639]

The challenge of the currently ongoing work is the further development of the now existing technical possibilities for a significant expansion of the direct by using the flat knitting technology realizable spectrum of net shape MLG.

Therefore the focus of actual developments is the creation of two new groups of MLG as semi-finished reinforcing structures. These are complex-shaped tubular textile structures with circumferentially encircling, uninterrupted reinforcing yarns (Figure 8 a, IGF 17926 BR) and secondly integral shell profile semi-finished products (Figure 8 b, IGF 18806 BR). Both semi- groups have an enormous potential for the development of highly complex FRP supporting structures based on net shape MLG.

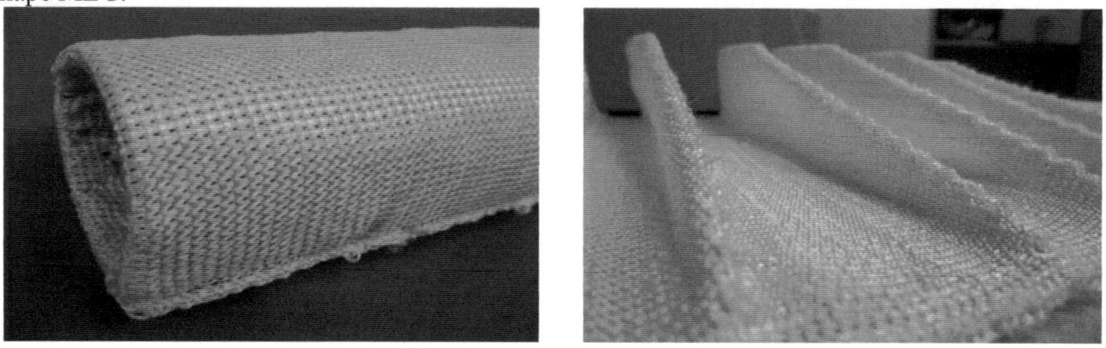

a b

Figure 8. Biaxial reinforced preforms a) tubular, b) shell-stringer geometry

3. Summary

Two enormous cost positions of FRP components are actually the production and the mostly manual preforming of textile reinforcing structures. The high costs limit currently a widespread use of FRP components. Therefore, approaches are urgently needed to allow a reduction of costs for the production of FRP-preforms and simultaneously create the prerequisites for an automated, high-volume compatible direct preforming.

A textile technology offering an enormous potential for achieving these goals while enabling a direct production of the 2D final contour or 3D net shape geometry of 2-layer biaxial reinforced structures is the flat knitting technique particularly the MLG technique. The samples impressively demonstrate that this technique offers a considerably simplified cost-effective and waste-free production of 2D or 3D preforms. The research results show that the number of reinforcing layers per MLG does not affect the mechanical properties of the composites produced therefrom. In addition, the shaping possibilities of the flat knitting technique can be used to arrange the reinforcing fibres according to the load direction.

The future challenge is the development of appropriate ways to directly achieve MLG with 2D final contour and / or 3D net shape geometry with more than two reinforcing layers and additional functionality [15].

4. Acknowledgment

The presented results show part of the research work at ITM within the collaborative research centre 639 (SFB639) for the realization of a complete process chain for the production of 3D biaxial reinforced MLG. The research work was funded by the Deutsche Forschungsgemeinschaft (German Research Foundation) (DFG 639, TP A2) at Technische Universität Dresden.

The IGF projects 17926 BR and 18806 BR of the research association Forschungskuratorium Textil e. V. are funded within the AiF program to promote joint industrial research (IGF) by the Federal Ministry of Economy and Energy, following a decision of the German Bundestag.

The authors gratefully acknowledge the financial support

5. References

[1] Homepage ShimaSeiki, http://www.shimaseiki.com/wholegarment/, (15.02.2016)

[2] Homepage Stoll, http://www.stoll.com/stoll-knit-and-wear-2/9_3, (15.02.2016)

[3] Hompage Nike, http://www.nike.com/de/de_de/c/innovation/flyknit, (15.02.2016)

[4] Haupt M, Lin H, Cherif C, Krzywinski S 2015 Weft-knitted preforms adapted for crash- and 3D-applications. In: 15th AUTEX World Textile Conference. Bucharest (Romania), June 10-12, 2015

[5] Cherif C 2016 Textile materials for lightweight constructions. Springer, Heidelberg, ISBN 978-3-662-46340-6.

[6] Schulz C 2013 Entwicklung einer Technologie für die Herstellung textiler Halbzeuge für thermoplastische Verbundwerkstoffe auf Basis der Mehrlagenflachstricktechnik, Dissertation, Dresden - ISBN 978-3-86780-337-3

[7] Orawattanasrikul S 2006 Experimentelle Analyse der Scherdeformationen biaxial verstärkter Mehrlagengestricke, Dissertation thesis, Universität Dresden

[8] Lässig R, Eisenhut M et al. 2012 Serienproduktion von hochfesten Faserverbundbauteilen: Perspektiven für den deutschen Maschinen- und Anlagenbau. Band 09

[9] Friedrich HE 2013 Leichtbau in der Fahrzeugtechnik. Imprint: Springer Vieweg, Wiesbaden, ISBN 3834821101.

[10] Eickenbusch H, Krauss O 2013 Kohlenstofffaserverstärkte Kunststoffe im Fahrzeugbau – Ressourceneffizienz und Technologie, VDI Zentrum Ressourceneffizienz, Berlin

[11] Trümper W 2015 Entwicklung einer unikalen Technologie für das direkte 2D- und 3D-Preforming von Mehrlagengestricken. Technische Universität Dresden, Fakultät Maschinenwesen, Institut für Textilmaschinen und Textile Hochleistungswerkstofftechnik, Dissertation, Dresden

[12] Lin H 2015 Entwicklung einer Prozesskette zur Umsetzung komplexer Geometrien in formgerechte 2D-3D Hybridgarn-Mehrlagengestricke für Faserkunststoffverbund-anwendungen. TUDpress, Dresden, ISBN 978-3-959080-156.

[13] Cherif C, Krzywinski S et al. 2012 Development of a process chain for the realization of multilayer weft knitted fabrics showing complex 2D/3D geometries for composite applications, *Textile Research Journal* **82**, 1195–1210

[14] Cherif C, Krzywinski S et al. 2013 New Process Chain for Realisation of Complex 2D/3D Weft Knitted Fabrics for Thermoplastic Composite Applications, *Procedia Materials Science* **2**, 111-129

[15] Cherif C 2015 Potential and challenges of multimaterial design for composites based on merging textile, metal and plastic technologies. In: 15th AUTEX World Textile Conference. Bucharest (Romania), June 10-12, 2015

Shape and Surface: The challenges and advantages of 3D techniques in innovative fashion, knitwear and product design

E Bendt
Niederrhein University of Applied Sciences, Faculty of Textile and Clothing Technology, Mönchengladbach, Germany

E-mail: ellen.bendt@hs-niederrhein.de

Abstract. The presentation wants to show what kind of problems fashion and textile designers are facing in 3D-knitwear design, especially regarding fashionable flat-knit styles, and how they can use different kinds of techniques and processes to generate new types of 3D-designs and structures. To create really new things we have to overcome standard development methods and traditional thinking and should start to open our minds again for the material itself to generate new advanced textile solutions.
This paper mainly introduces different results of research projects worked out in the master program "Textile Produkte" during lectures in "Innovative Product Design" and "Experimental Knitting".

1. Introduction

With regard to knitted fabric – especially knitted products - 3D techniques have been an integral part of the development process since the very first beginnings of textile production. The creative process for the designer starts with the yarn, unlike in textile products which refer to already existing textile fabric. This especially concerns the intersections of fashion- and textile design. Digitalization and modern CAD applications made the development and design of many products easier; they lead, however, to stringency in product development, which allows for only little or no creative encounter or involvement of the designer in the technical development process. This is then inevitably – just like with many textile products – no longer any kind of fundamental re-development but merely an update or a variant of an already existing one.

All fashion and textile designers in creative industries are familiar with graphic programs like Illustrator or Photoshop, but only few have some knowledge of knitting techniques, which are needed to create new ideas for fabric, shape and surfaces in knitwear. So the results and developments are dependent on a multitude of factors like the interpretation of the technician. That circumstance bears various risks with regard to the realisation and outcome of a design, its final look, and it makes the design not always satisfying for the designer neither the technician.

2. Challenges and Advantages of the different techniques: Seamless Technology for flat knit designs?

Especially the seamless technology for fashionable flat knit styles offers several problems for the development of new ideas. On the new X-bed machines it is possible to work out knits in double or multiple layer techniques, but the machines are expensive, not always available and not easy to handle.

Furthermore, an experienced knitting technician is required to do the programming for the machines and to let them run. So, as mentioned above, in most cases the design is not only made by the designer, but in cooperation with a technician who sometimes is thousands of miles away in an out-sourced place of production.

The narrowings and fashion marks needed in the machine programs and their position in the style panel are, on the one hand, part of the aesthetic appearance of a product and, on the other hand, very important for the correct fitting and shape. This is a big challenge for designers of fashionable products who have to combine these aspects with trendy yarns, patterns and shapes, because there is a lack of training programs and communication tools (like model libraries). This circumstance often leads to a gap in communication, and battles with the technical department about possibilities and frustration with the outcome of ideas.

Nevertheless, more and more advantages of the complete garment production like minimum waste yarn and post-operations, lean logistics, the possibility of European production and other sustainable aspects are focused by industry, and promise a growing potential in the future, adding to the increasing importance of the fully fashion production.

If design skills and technical know-how are perfectly connected, generating products with fully fashioned technique and whole garment knitting can lead to beautiful and perfect fitting, seamless shapes and structures (Fig. 1).

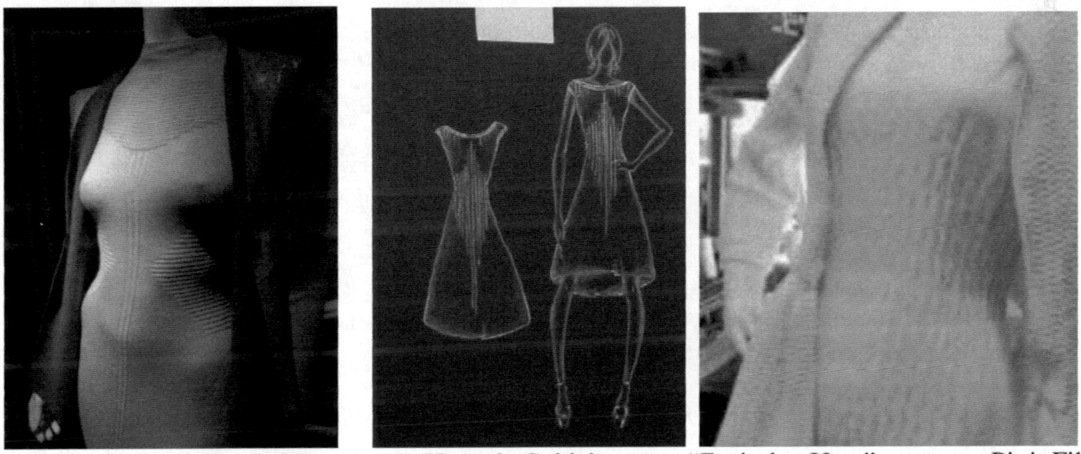

Figure 1. Left: Fully fashion dress, Hannah Seidelmeyer, "Feel the Yarn" contest, Pitti Filati, Florence, 2013; middle and right: fully fashion dress, Marina Keberlein, "0°", "Wool School" contest 2012

Strategies to start creative processes: Let us start with the yarn

To introduce knitwear design for weft-knit styles to students, it is better not to start with a sketch but with the yarn and a material concept on hand-knitting machines or by hand, to get access to the material and "other" ways of creative developments. The goal is not only to make simple plain fabric, but to play with stitches, constructions and structures, yarn combinations, tensions etc. to get a feeling for knitwear and to discover new design strategies. "Trial and error" and results from overcoming obstacles are an important part of design development.

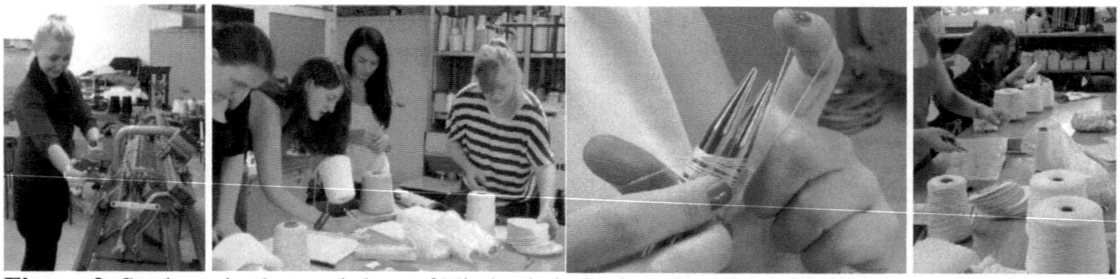

Figure 2. Students in the workshop of Niederrhein University of Applied Sciences.

Things to be discovered by the students, during lessons and workshops:

For the shape of a style panel seams und cutting of materials can be avoided by combining different stitch and fabric structures, yarns etc. even if narrowings and fashion marks are not used, like in the fully fashion technique (Fig. 3).

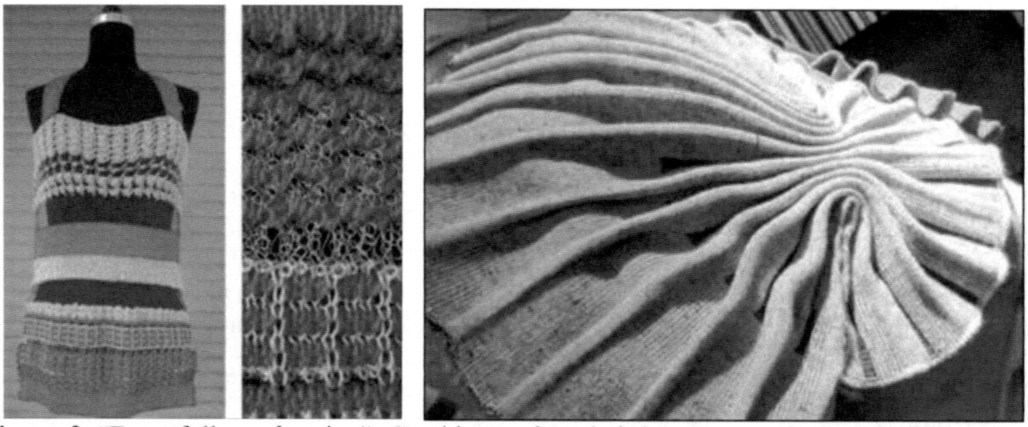

Figure 3. "Form follows function": Combing various knitting constructions with different properties, Sarah Stober, 2012 and knitted pleats (left picture: Stober 2012).

To use small models (e.g. size 1:4) to test various proportions, material combinations and draping effects of test swatches (Fig. 4).

Figure 4. Artworks of students for the project "Knitted Wool Couture", 2012 (picture: Windgassen/2012)

There are a lot of variables in the knitting process which can lead to unexpected 3D innovations: different yarn types and subtleties, strength parameters, yarn tension, a coincidentally "wrongly" distributed thread guide or an error in programming, lead to movable surfaces or other astonishing effects as shown in the following projects.

Case study: Experimental flat knitting with an "unusual" yarn: Diolen 150 BT, by PHP Fibers GmbH (sustainable Polyester)

For the "Apparel Show" at the fair "Techtextil/Texprocess" in Frankfurt/Germany in May 2015, the goal for the participants was to show that typical technical materials can be used for clothing as well. An interesting sustainable material was found to work with: Diolen 150 BT, a biodegradable high-tenacity polyester on PLA basis, made from renewable resources of company PHP Fibers, Germany. It is a shiny white yarn for warp knitting, actually used in the automotive interior for safety belts and air bags and not for clothing. It was perfect for the project. The task for the students was to create something new with it. They were not asked to make simple plain fabric, but to work with the properties of the material, like stiffness, to implement it in a clothing concept.

The project team started with some knitting tests by hand, on hand-driven and computerized (Stoll CMS 302 TC) flat-knitting machines. Because of the harsh handfeel of this polyester type it took some time until the results were satisfying. The students "played" with tension, knitting constructions, tested different finishings and combinations with wool or other materials. As expected before: in the end some very different surprising results with 3D effects appeared:

"La Onda del Hielo – Moving surfaces"

Master student Sophia Krinner did tests by using Diolen in combination with tuck stitch ribs in a loose knitting tension. The resulting fabric showed a kind of chain reaction on the surface if tugged at one end. The ribs started moving and the whole surface changes. The student felt inspired to use this special property of the material to develop the couture dress "La Onda del Hielo", a dress with a moving surface while wearing it (Fig. 5).

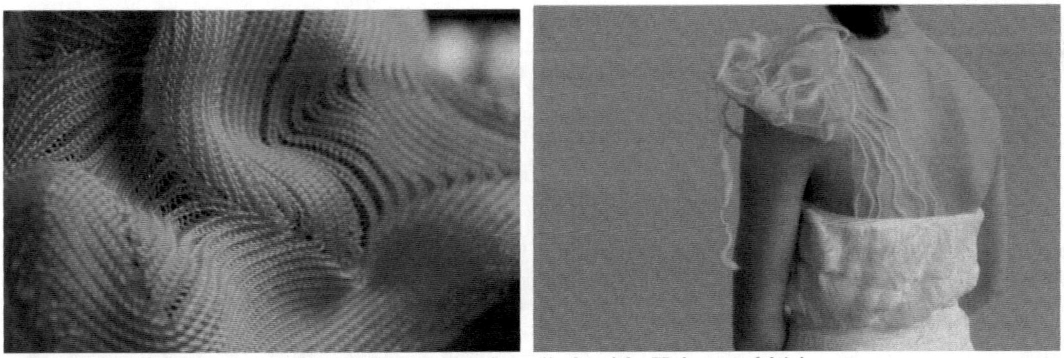

Figure 5. "La Onda del Hielo – Moving surfaces", Sophia Krinner, 2014.

"Fading Crystals – Ultrasonic welding for flat knit styles"

Another partially surprising result occurs, when new joining technology, as for example in this case ultrasonic welding, on the surface of coarse knitted clothing or other – yet untypical – materials is used. Master student Theresa Brinkmann, did experiments with structures on the surface of different stitch constructions, pleats and pleat optics and finally received special pleats with a look like knitted "petinet" effect by using the right emboss tool and correct parameters like pressure and timing on the welding machine. The result: a couture dress with a voluminous 3D-bottom part (Fig. 6).

Figure 6. "Fading Crystals", Theresa Brinkmann, 2015 (pictures. left Brinkmann 2015, middle and right Ilskens 2015).

Case study: Experimental flat knitting with wool combining yarns with different subtleties and using of felting techniques to gain volume

Other student projects dealt with structures and volumes, by using different yarn types in one style, and/or the possibility of felting wool in different ways:

"Moonbotica":

Following the semester topic "metamorphosis" master student Verena Winkelmann created the knit project "Moonbotica", outlining the development from cocoon to silk moth as a chunky sweater with "growing" stitch sizes from the bottom to the top. The pattern is 100 % wool hand knitted and hand felted. The felted bottom part characterizes the soft but compact cocoon. Towards the upper part the structure breaks open and spreads in a heavy chunky knit to distinguish the detailed pattern of the silk moth's wings [1]. For this part the student used woollen tops instead of yarn (Fig. 7).

Figure 7. "Moonbotica", Verena Winkelmann, 2013 (right picture: Berns/Winkelmann 2014)

"Knitted Wool Couture" Wool School - Cooperation Project for the "Campaign for Wool" in 2012 - project partner "Woolmark International" and fashion brand "Marc Cain":

"The Wool School Project" was part of "The Campaign for Wool" in 2012 that is a global endeavour initiated by its patron, His Royal Highness The Prince of Wales, in order to raise awareness amongst consumers about the unique, natural and sustainable benefits offered by wool. The goal for the HSN students: To develope something different to what people would expect and to show a much more-in-depth involvement with the material than in classical fashion design. Volume and many different 3D-surfaces and structures appeared in untypical yarn combinations and felting techniques like needle punching and the usage of felt yarns, to create "Knitted Wool Couture" (Figs. 8, 9).

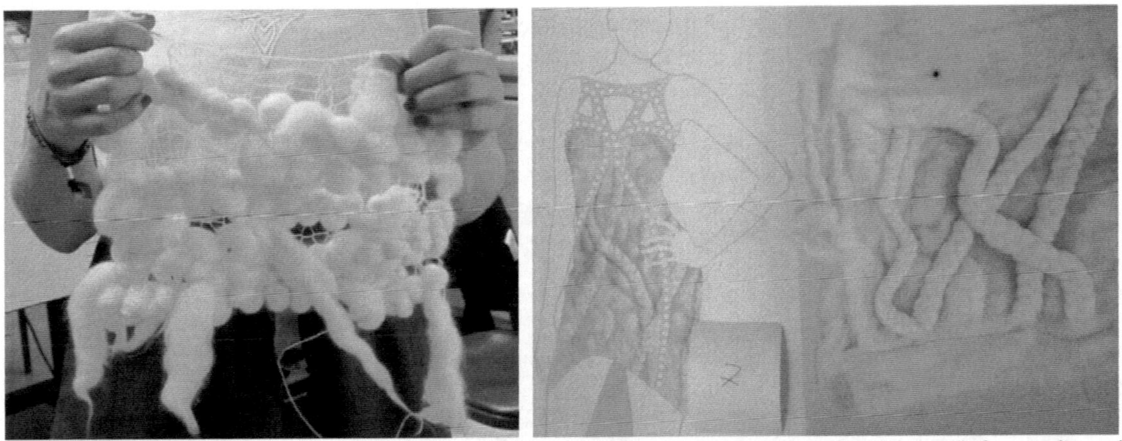

Figure 8. Combination of an Nm 30/2 yarn with woollen tops by Caroline Liehr (left panel) and a felting structure in different subtleties and tops by Julia Tschukin (right panel).

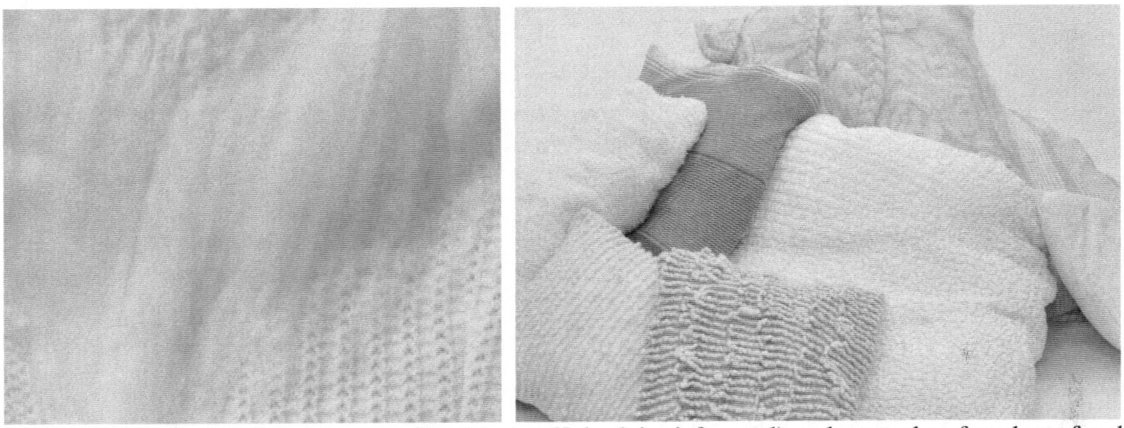

Figure 9. Needle punching structure (by Marina Keberlein, left panel) and artworks of students for the project "Knitted Wool Couture", 2012 (right panel) (pictures: Bendt & Windgassen/2012).

The winning outfit in this project and contest impressed the jury with its architectonic shape and volume (Fig. 10).

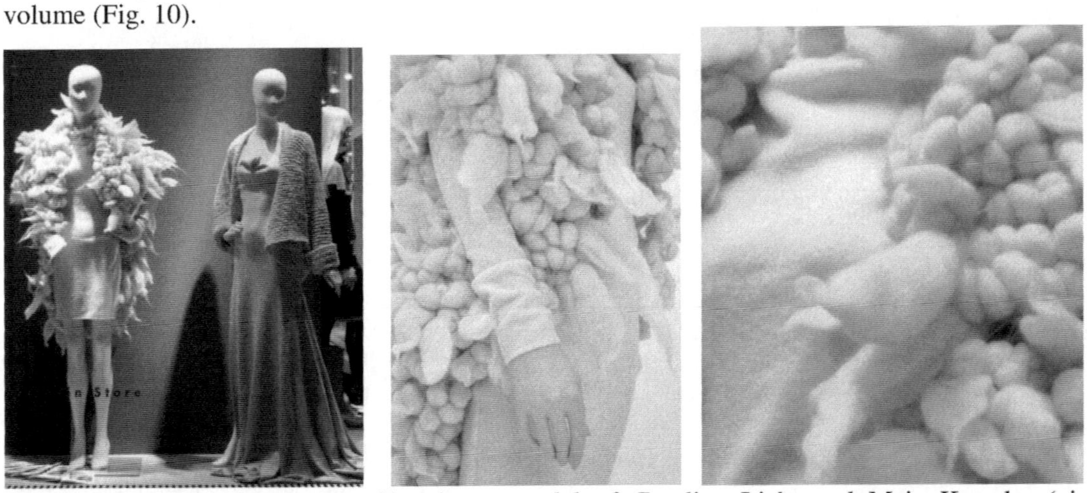

Figure 10. "Crazy, Sexy, Wool", winner model of Caroline Liehr and Maia Kesseler (pictures Shabati, Windgassen and Bendt 2012)

Results of another master research project, exploring the potential and outcome of knitting felt yarns and non-felting yarns together in one pattern, show interesting, structured "cloqué" effects, after finishing/washing processes, e.g. in jacquards with various backings (Fig. 11).

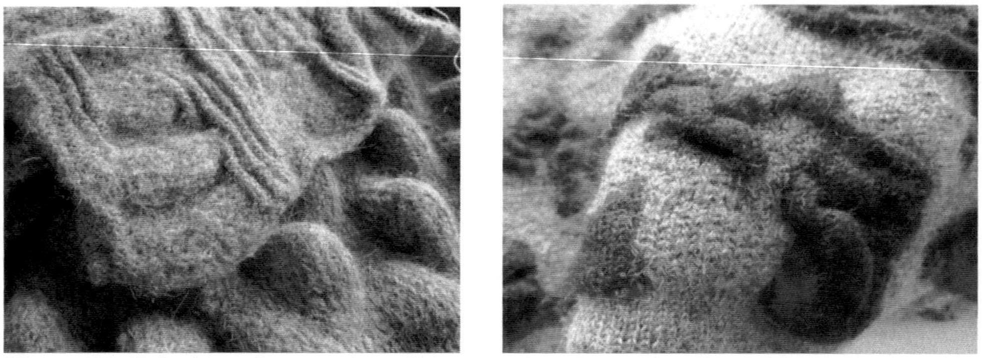

Figure 11. Felted woolen samples of master research project (Sarah Grobe).

To start with the fabric:
Warp knitted lace meets silicon - a basis for a new textile 3D product design:

The possibility to use different open stitch structures of warp knitted lace for a new type of 3D-textile material in the shoe sector was developed by master student Ariane Ehring with the project "Grow" in the course "Innovative Product Design". It is inspired by bionic structures as can be found in "*coral reefs, stone reliefs and on the skin of reptiles. Natural textures are imitated with artificial and synthetic materials. For the realisation of the collection different coloured silicon and laces have been developed. Due to the variations in the open structure of the lace fabric it is possible to create multiple surfaces*" [2] (Fig. 12).

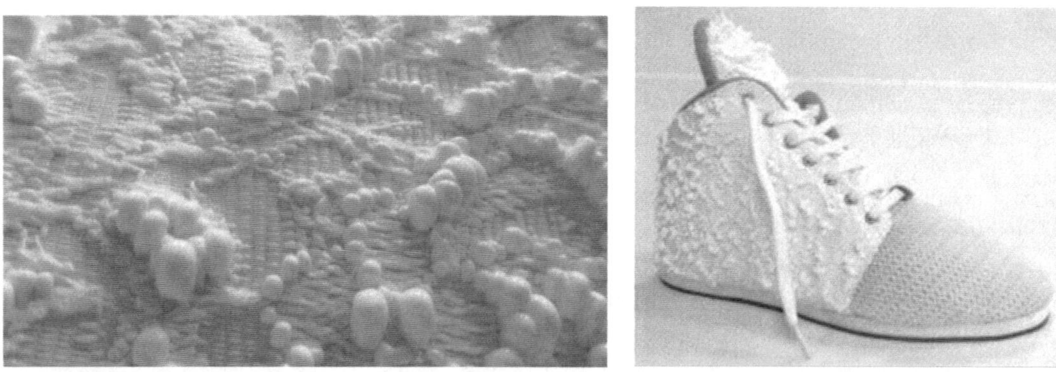

Figure 12. "Grow" by Ariane Ehring, 2014.

3. Conclusion

Nowadays knitting technology offers an amazing range of possibilities for designing including 3D-materials and styles. Unfortunately, only little of it is used in the textile and clothing industry. One problem master student Miya Budaeva found out in her master research project, on the usage of seamless technology in the fashionable flat knitting sector was that product developers in industry, fashion and textile designer do not have enough technical knowledge to work with and that they are not aware of the advantages and possibilities of whole garment technologies.

The second problem is that under consideration of cost saving aspects, production- and development facilities – especially in the fashion industry – have increasingly been globally re-located and standardised during the last 30 years. Hence, the ideas-generating designers can very often no longer accompany the creative developing process together with the technicians that do the machine programs. Especially in knitwear this is a problem because really new knitwear developments cannot be generated without a more-in-depth involvement with the material, knitting and finishing tests and creative guidance.

So, to change something for the future and to generate more innovative designs and materials, there should be more technical training and exercises with materials during the education and studies of designers and product developers at universities. And this means not only from the vantage point of fashion. Knitwear offers comfort, flexibility, seamless shapes and fabric constructions with a lot of functional potential like, e.g., the plating technique for applications in sports and work wear or for medical and protection garments.

The development of creative tools like pattern libraries, maybe connected to standard design programs, should be promoted and the textile- and creative industries should be more aware of the advantages of seamless production – like, e.g. sustainable aspects. This offers the opportunities to a creative, sustainable and innovative 3D-knit design world.

References

[1] Winkelmann V 2015 The German Look at Design – Advanced Textile Solutions, 1st edition, p. 80, ed. Bendt/Wachs, Hochschule Niederrhein, Mönchengladbach

[2] Ehring A 2015 The German Look at Design – Advanced Textile Solutions, 1st edition, p. 130, ed. Bendt/Wachs, Hochschule Niederrhein, Mönchengladbach

IOP Conf. Series: Materials Science and Engineering **141** (2016) 012017 doi:10.1088/1757-899X/141/1/012017

Structure and Elongation of fine Ladies' Hosiery

M Lozo[1] and Z Vrljicak[2]
[1]Tvornica carapa, 8.mart, Subotica, Serbia
[2]Tekstilno-tehnoloski fakultet, Zagreb, Croatia

E-mail: 8mart@eunet.rs and krivodol@yahoo.com

Abstract. On a sock-knitting machine with diameter of cylindrical needle bed 100 mm (4e″) that knitted with 400 needles, samples of fine women's hosiery were made from four PA filament yarns in counts 20 dtex f 20, 30 dtex f 34, 40 dtex f 40 and 60 dtex f 60. Each type of yarns was used to make hosiery samples with four loop sinking depths of unit values in a computer program 400, 550, 700 and 850. For all the samples, parameters of yarn structure were analyzed and elongation properties of knitted fabric were measured. During the elongation of knitted fabric, areas of knitted fabric elasticity, beginning of permanent deformation and elongation at break were measured. Elongation of knitted fabric in the wale direction, i.e. transverse hosiery elongation and elongation of knitted fabric in the course direction, or longitudinal direction of hosiery were measured. Yarn fineness and loop sinking depth significantly influence the elongation properties of hosiery.

1. Introduction

Fine ladies' hosiery is made on specially designed sock-knitting machines. Those are cylindrical machines of small-diameter needle bed, most commonly 90 to 125 mm (3 ½ to 5e″). Hosiery is basically intended for adult women and is often made on automatic machines with cylinder diameter of 100 mm (4e″), that use 400 needles, i.e. have the gauges E32, cf. Table 1. On such an automatic machine, a sock is knitted in 120 to 300 s, depending on the constructional shape of a sock and a series of technological parameters. After the knitting process, the sock needs to be finished in the sewing department. In case of pantyhose, two legs are joined, without or with different inlays. The sewing process is followed by the refinement process, including sock dyeing and ironing [1-3].

Table 1. Relation between cylinder diameter, machine gauge and number of needles in one-cylinder automatic sock-knitting machines designed for making fine ladies' hosiery [4]

Machine gauge, E	Number of needles in a machine					
	Cylinder diameter, (approximate)					
	mm → 90	95	100	110	115	125
	in → 3 ½	3 ¾	4	4 ¼	4 ½	5
24	260/264	284	304			376
26	280		328			408
28	306	330	340	376	396	440
30	330	354	376	400	424	472
32	352	376	400/402	432	452	500
34	370	400	422		480	
36		424				
38		448		512		

Fine women's hosiery is usually made using three types of thread [5-7]:
1. polyamide (PA) filament threads,
2. polyester (PES) filament threads,
3. elastane threads.

The given thread types are used in various combinations depending on the purpose of hosiery. In the production of simpler and cheaper hosiery, only one thread composition is used, usually only PA filament threads, rarely PES threads, Figure 1. Apart from PA filament thread, different elastane threads (e.g. Lycra) are used due to hosiery's comfortable fit on the leg and wearing comfort. Elastane thread can be knitted into each row next to the PA filament thread, or knitted only into certain rows or separate sections of hosiery. When designing structure and tensile properties of hosiery at its particular sections, a designer will use the elongation properties of PA and elastane yarns and incorporate them to an extent into certain parts of hosiery. The share of elastane thread significantly determines hosiery's elongation properties, i.e. its compressibility on the leg, and thereby its wearing comfort.

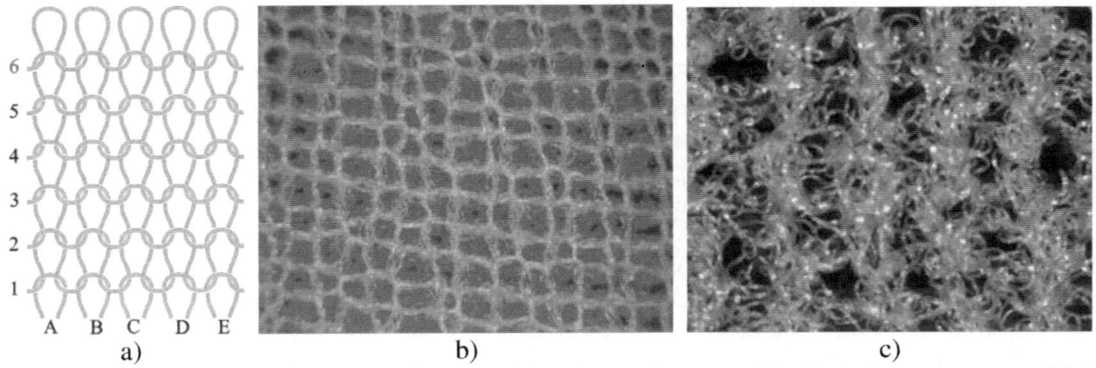

Figure 1. Basic structure of simple fine women's PA hosiery; a) plain is the base structure of ladies' hosiery, b) position of PA filament thread in the basic structure of hosiery – a stretched structure and c) position of threads in shrunk hosiery.

One of hosiery's roles is to protect the leg from external influences, primarily cold and wind. Therefore, the structure of filament yarns for hosiery production is different than the structure of filament yarns for the production of clothes: T-shirt, shirts, dresses, skirts, pants, etc.

Filament yarns for hosiery production are made of several fibers. Fineness and number of fibers determine the fineness of filament yarn. Number of twists determines the structure of filament yarn. Greater number of threads in filament yarn brings greater wearing comfort of hosiery. Yarn with a larger number of fibers produces more difficulties and hold-ups in the knitting process causing a significantly higher amount of errors, losses and production waste. It is necessary to find an optimal number of fibers that will make-up a filament yarn according to the product's purpose. On the market, there are various counts of filament yarns with different number of fibers. Filament yarns from 10 to 33 dtex, most commonly 13, 17, 22 and 33 dtex are usually used to make fine women's hosiery. These yarns may contain 5 to 37 fibers which usually have a circular cross-section and counts in 2.5 to 3.5 dtex. Somewhat coarser hosiery is made of 20 to 60 dtex yarns, most commonly 20, 33, 40 and 60 dtex with microfibers from 0.8 to 1 dtex [8,9].

There are a few methods of making filament yarns for hosiery production. Classic filament yarns are made with very few twists, usually up to 50 twists/m, or without twists, i.e. fibers are not twisted during yarn construction. In this case, yarn in the knitted fabric is not compact but spread out on the surface it covers. In that way, fibers have a bigger surface of fit on the leg and produce a higher surface friction coefficient and greater wearing comfort. Only much finer fibers are used in the construction of the other yarn shape, often 1.5 to 2.5 dtex, of which there are more in a single yarn. Use of microfibers results in high-quality hosiery. In order to prevent fibers from causing problems during knitting, yarns are ostensibly stuck together at every 10, 15, 20 or 25 mm and thus knitted into loop wales. Elongation at break of PA filament yarns for hosiery production is 20 to 35 %, and that of elastane yarns is 200 to 900 %. With such yarns, it is possible to accomplish different extensibility and compressibility of hosiery on the leg. In the production of classic cotton short socks, cotton yarns of elongation at break around 5 % are used, cf. Figure 2.

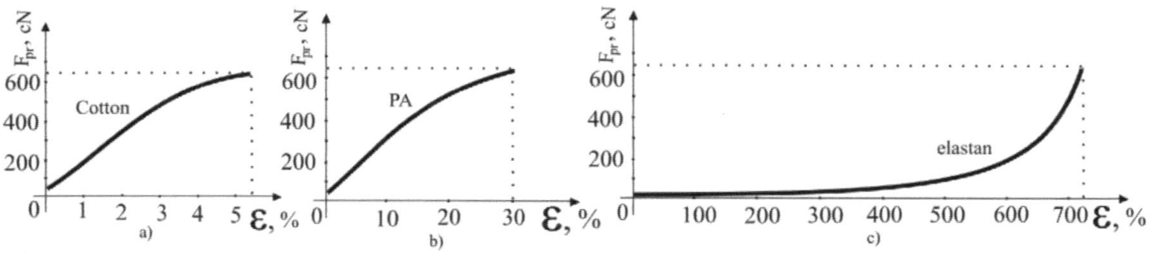

Figure 2. Yarn elongation properties in hosiery production: a) cotton yarns used in the production of short socks, b) PA filament yarns for the production of fine ladies' hosiery and c) elastane yarns

When designing fine ladies' hosiery a constructor or designer must be familiar with measures of the leg for which the hosiery is intended and fit the measures to the sizes found in the standard. In order to make a functional product, a designer should also be familiar with structures, and primarily elongation properties of yarns they want to use to make fine hosiery. Fibers that build the yarn can be smooth or textured, i.e. slightly wavy. Some yarn structures are used to make hosiery for particular purposes.

Structure and elongation properties of yarns are key parameters in hosiery design. Yarn properties should be consolidated with constructional properties of a machine in order to make hosiery of a particular size and wearing comfort. In different interlacements, a single sock-knitting machine can be used to knit with filament yarns of certain yarn count, structure and elongation properties. When making hosiery using a single count of filament yarn, it is necessary to obtain certain widths of a knit on different hosiery sections that will envelop the leg comfortably. Apart from different widths in one sock, it is necessary to obtain a certain hosiery elongation. This is usually realized by the application a certain interlacement and regulation of loop sinking depth on the machine. Therefore, yarn count and elongation properties of filament yarn, as well as the type of interlacement and loop sinking depth are the main parameters that determine elongation properties of hosiery being used. That is why the experimental part of this paper investigates the influence of yarn count and loop sinking depth on elongation properties of fine women's hosiery.

2. Experimental work

The aim of this research is to get information about the extent of influence of count of PA filament yarn and loop sinking depth on the structure and elongation properties of fine women's hosiery. A machine used in everyday production was used to make sample hosiery. Simpler forms of fine women's hosiery available on the market are made in plain structure, cf. Figure 1a. This interlacement was used to conduct this experiment. Hosiery is produced from filament PA yarns of different counts and structures. Four different kinds of PA yarn counts and four loop sinking depths were used in this experiment.

3. Properties of the machine for the production of samples

For this research, an automatic sock-knitting machine with diameter of cylindrical needle bed 4e" that knitted with 400 needles on four knitting systems was used, Table 2. The distance between the centers of two neighbouring needles is 0.80 mm. Needle body thickness is 0.33 mm, and the hook for picking up threads 0.26 mm. In this construction of automatic machine, needle hook thickness is 2 to 5 times bigger than the conditional knitting-yarn thickness, i.e. such automatic machines are recommended for producing fine women's hosiery with threads of conditional thickness 0.05 to 0.1 mm, i.e. yarn count in 13 to 72 dtex. A single type of yarn and different loop sinking depths are used to make fine women's hosiery. In this case, sampling hosiery and measuring their elongation is difficult, because two or three different yarn densities obtained at particular loop sinking depths are mixed on a small surface of the knitted fabric. Therefore, the hosiery samples used in this research were made at a single loop sinking depth. Samples of tubular knits and socks are 40 to 80 cm long and suitable for various analyses.

Table 2. Properties of the machine used for the production of fine women's hosiery samples

Machine gauge, E	Cylinder diameter mm	e"	Number of knitting systems, S	Number of needles, N_i	Cylinder operating speed, min^{-1}
32	100	4	4	400	250 do 700

4. Properties of PA filament yarns for the production of samples

PA filament yarns of four different types of fineness were used. Every yarn has its properties, Table 3. The finest yarn has 20 dtex, and the coarsest 60 dtex. Yarn in 20 dtex is made with 20 fibrils and yarn in 60 dtex with 60 fibrils. Predominant fineness of fibrils is 1 dtex. Yarn elongation properties are presented in Figure 3 and Table 4. In diagrams showing yarn elongation properties it can be seen that yarns are uneven and that some fibrils are broken. Yarns in 40 dtex f 40 have the smallest elongation at break of 21.4±0.4 %, while yarns in 20 dtex f 20 have 25.9 ± 0.4 %. Breaking loads are within limits 3.8 to 5.0 cN/dtex. The given properties of yarns are the basic parameters used in theoretical research of relation between yarn elongation properties and hosiery united in an interlacement.

Table 3. Properties of PA fibrils and yarns for the production of samples

Loop sinking depth →			400	550	700	850
Yarn count	Approximate fibril diameter, mm	Approximate total surface of fibril cross-section, mm^2	Sample label			
20 dtex f 20	0.012	0.00226	1 X C	1 X Z	1 X P	1 X O
33 dtex f 34	0.011	0.00322	2 X C	2 X Z	2 X P	2 X O
40 dtex f 40	0.012	0.00452	3 X C	3 X Z	3 X P	3 X O
60 dtex f 60	0.012	0.00678	4 X C	4 X Z	4 X P	4 X O

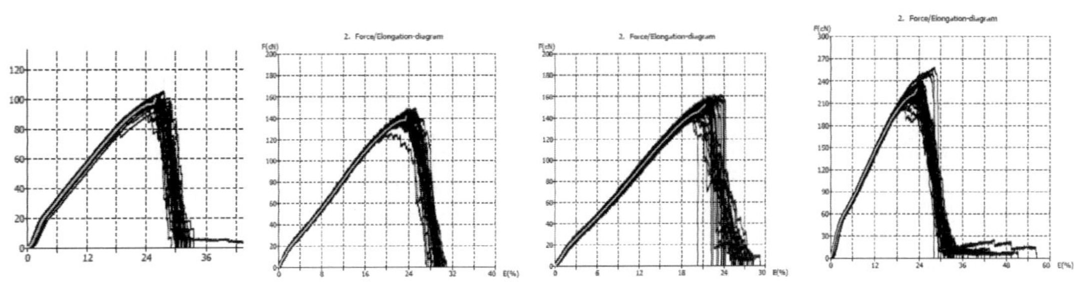

a) b) c) d)

Figure 3. Diagrams of elongation properties of polyamide filament yarns for the production of samples: a) 20 dtex f 20, b) 33 dtex f 34, c) 40 dtex f 40 and d) 60 dtex f 60

Four knit samples at different loop sinking depths were made using each yarn. The machine works in CAD/CAM system whose smallest loop sinking depth is 400 and biggest 850 units. Samples made at the smallest loop sinking depth are labelled with C, and those at the biggest with O, Table 3.

Table 4. Properties of polyamide filament yarns for the production of hosiery samples

Yarn properties	**Samples**			
	20	**33**	**40**	**60**
Yarn count, dtex	20.1±0.2	33.1±0.2	40.2±0.3	60.0±0.2
Breaking force, cN	100±1	142±1	153±2	230±4
Elongation at break, %	25.9±0.4	23.5±0.3	21.4±0.4	23.2±0.6
Work to break, cN·cm	752±17	931±24	864±29	1581±70
Breaking load, cN/dtex	5.0	4.3	3.8	3.8

5. Results of measuring parameters of knit structure in fine women's hosiery

Samples of fine women's hosiery were knitted using four PA filament yarns in counts 20, 33, 40 and 60 dtex, Table 5. Each yarn was used to make four basic samples. Each sample was made with a different loop sinking depth of 400, 550, 700 and 850 units (orientationally nondimensional number in a computer programme) [10]. Therefore, four groups of samples in four yarn counts were produced, or the total of 16 basic samples. Each sample was used to make ten pieces of hosiery, i.e. the total amount of hosiery knitted was 160 pieces, which were all torn during the analysis of the knit and measurement of yarn elongation properties.

Table 5. Parameters of knit structure in fine women's hosiery produced

T_t dtex	h_k	m, g/m^2	S_p, mm	d_{pl}, mm	m_z, g/cm^3	D_h, l./cm	D_v, l./cm	C	D, l./cm^2	ℓ, mm	L, mm/cm^2
20	**400**	38	118	0.28	0.146	17.0	45.9	0.59	779	2.16	1684
20	**550**	37	117	0.31	0.126	17.1	37.6	0.60	642	2.49	1598
20	**700**	35	122	0.33	0.112	16.4	33.4	0.77	548	2.80	1535
20	**850**	34	122	0.32	0.113	16.3	29.1	0.77	475	3.13	1486
33	**400**	55	111	0.32	0.184	18.1	40.3	0.46	730	2.19	1598
33	**550**	51	111	0.34	0.159	18.1	32.0	0.56	577	2.53	1460
33	**700**	50	124	0.37	0.141	16.1	30.7	0.67	494	2.86	1413
33	**850**	50	123	0.40	0.130	16.3	27.3	0.78	444	3.20	1422
40	**400**	58	118	0.32	0.191	17.0	38.3	0.44	651	2.17	1413
40	**550**	54	120	0.35	0.163	16.7	31.2	0.54	521	2.52	1312
40	**700**	50	128	0.37	0.143	15.6	27.1	0.76	422	2.81	1186
40	**850**	49	133	0.41	0.127	15.0	24.7	0.81	371	3.21	1191
60	**400**	99	116	0.38	0.276	17.2	43.3	0.34	744	2.15	1599
60	**550**	92	116	0.43	0.223	17.2	34.2	0.46	589	2.47	1455
60	**700**	83	119	0.45	0.193	16.8	28.1	0.58	471	2.82	1328
60	**850**	77	130	0.44	0.182	15.3	25.3	0.72	388	3.17	1229

Abbreviations: T_t – yarn count, dtex, h_k – loop sinking depth, dimensionless number, m – yarn mass per unit area, g/m^2, S_p – knit width, mm, d_{pl} – knit thickness, mm, m_z – knit volume mass, g/cm^3, D_h – compactness of loops in the knit wale, l./cm, D_v – compactness of loops in the knit course, l./cm, C – coefficient of loop compactness, D – number of loops per basic unit of surface, l./cm^2, ℓ - thread consumption in a loop, mm, L – thread consumption per basic unit of surface, mm/cm^2

Diameter of cylindrical needle bed of the automatic machine used to produce hosiery is 100 mm and its perimeter is around 315 mm. On a cylinder of this size, different lengths of hosiery were produced. The narrowest sock has the width of 111 mm x 2 (tubular shape) and it is produced by knitting yarns in 33 dtex and loop sinking depth of 400. In this sock, shrinkage of the knit in the wale direction after removal from the machine and relaxation is 30 %. The widest sock is obtained by knitting with the coarsest yarn, i.e. yarn in count 60 dtex and loop sinking depth of 850, and it is 133 mm x 2. In this sock, knit shrinkage is around 15 %. Knit mass per unit area for hosiery is within the range of 34 and 99 g/m^2, and knit thickness in hosiery between 0.28 and 0.45 mm. Information about the knit given refers to one loop so as to demonstrate the fineness of the knit structure. Thread consumption for the formation of one loop ranges from 2.15 to 3.21 mm where the knit has 371 to 779 l/cm^2, or 1186 to 1648 mm of yarn is woven into one cm^2 of the knit. These knits are currently considered to be among finest knits in the world. An operating analyst striving to analyze parameters of their structure must be very experienced in order to conduct a proper analysis.

6. Results of measuring hosiery elongation

When using hosiery, the most important is the extensibility or elongation when tensile force is applied in the wale direction, or the so called transverse elongation, Fig. 4 and Table 6. It is significant when a sock is being put on a leg. Hosiery elongation in the course direction or the so called elongation along the sock is equally important, Table 7. It is primarily related to the height of a person wearing hosiery, as well as hosiery size. However, an important factor during long-term wear is the orbicular elongation of a sock which is flatly pressing the leg, and represents an important parameter in the evaluation of hosiery's wearing comfort. Orbicular elongation of a sock is the basis for determining its compressibility. This parameter is significant for the use of numerous hosiery types, e.g. socks for special uses: for soldiers and athletes (football, handball and volleyball players, skiers, etc.), for therapeutic purposes (injured patients who are often in a comatose condition), medical applications (prevention or protection of varicose veins) [11,12].

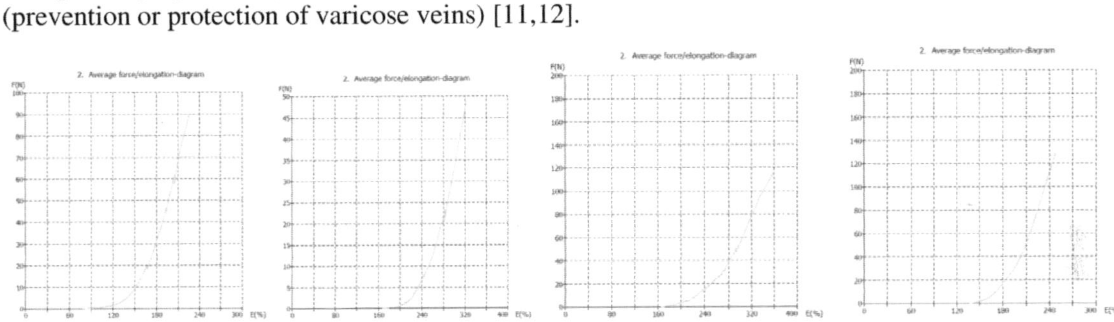

a) b) c) d)

Figure 4. Diagrams of elongation properties of knits in fine women's hosiery: a) yarn count 20 dtex f 20, loop sinking depth 400, transverse direction, b) yarn count 20 dtex f 20, loop sinking depth 400, longitudinal direction, c) yarn count 60 dtex f 60, loop sinking depth 850, transverse direction d) yarn count 60 dtex f 60, loop sinking depth 850, longitudinal direction

In this research, hosiery elongation was measured in the wale direction and course direction. Samples were 50 mm wide and 150 mm long for all measurements while the distance between clamps on the testing machine was 100 mm. Figure 4 shows mean tear values of several main knit samples. Three important values related to hosiery elongation were read from the diagram [13]. First, the linear part of the diagram to the point T_1 represents an elastic area (ε_e) which in structures like this occupies around 53 % of total knit elongation, Figure 5. The other part of the diagram from the point T_1 to the point T_2 represents a boundary area or a part that connects the elastic area and the area of permanent deformation. Its share in this kind of structure is around 14 %. The third part of the diagram stretches from the point T_2 to the point of break and represents permanent or plastic deformation (ε_p). This part makes up around 33 % of total elongation. During elongation and in the moment of knit tearing, the testing machine records the length of elongation at break (ε_t), which is usually not controversial. However, simple and accurate determination of the area of elasticity, i.e. point T_1 and the beginning of permanent deformation, i.e. point T_2 is not always easy. In different knit structures including fine women's hosiery, this area, i.e. the boundary area of elasticity and the area where permanent deformation begins, has not been investigated enough. Table 6 and Table 7 provide the values that were estimated for the given points.

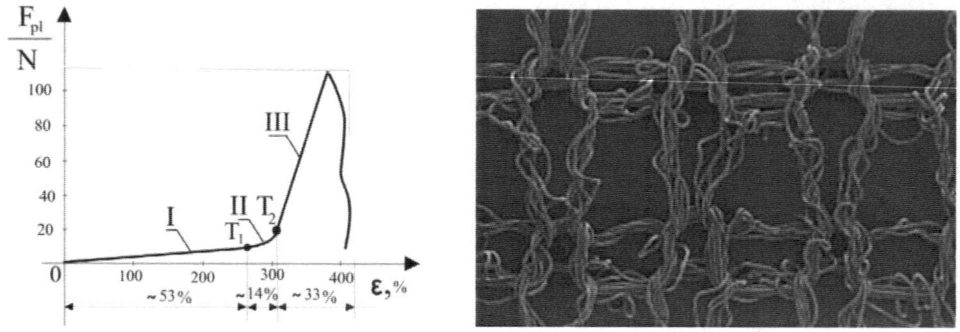

a) b)

Figure 5. Hosiery elongation; a) basic force/elongation diagram for fine women's hosiery and b) increased structure of the elongated knit with visible positions of some poorly spun fibers in filament yarn.

Table 6. Results of knit elongation in fine women's hosiery in the wale direction – transversely

Yarn count	Loop sinking depth	ε_{ep}, %	$\Delta\varepsilon_{ep}$, %	ε_{pp}, %	$\Delta\varepsilon_{pp}$, %	$\Delta\varepsilon_{pp}$ - $\Delta\varepsilon_{ep}$ %	ε_{tp}, %	$\Delta\varepsilon_{tp}$ - $\Delta\varepsilon_{pp}$ %
20	**400**	90	40	150	66	26	227	34
20	**550**	180	60	220	73	13	302	27
20	**700**	260	67	290	75	8	389	25
20	**850**	250	68	270	74	5	366	26
33	**400**	140	56	160	64	8	249	36
33	**550**	190	58	220	67	9	330	33
33	**700**	220	64	240	70	6	343	30
33	**850**	200	53	260	69	16	378	31
40	**400**	90	45	130	65	20	200	35
40	**550**	120	45	170	64	19	267	36
40	**700**	160	52	210	68	16	310	32
40	**850**	200	57	240	69	11	349	31
60	**400**	60	30	110	55	25	200	45
60	**550**	120	44	160	59	15	270	41
60	**700**	180	53	240	70	18	342	30
60	**850**	180	50	250	69	19	360	31
			53±5		67±3	14±3		33±2

Abbreviations: ε_e – elongation or extensibility of the knit to the point T_1, elastic area, %; ε_p – elongation or extensibility of the knit to the point T_2, to the beginning of the area of elasticity, %; ε_t – elongation or extensibility of the knit to the moment of break, %, $\Delta\varepsilon_e$ - share of elastic area in relation to total elongation, %; $\Delta\varepsilon_p$ - share to the beginning of plastic area in relation to total elongation, %; $\Delta\varepsilon_p - \Delta\varepsilon_e$ - share between the points T_1 and T_2, %; $\Delta\varepsilon_t - \Delta\varepsilon_p$ - share between the point of break and T_2, % ; index *p* marks the wale direction – transversely, and in Table 7, longitudinal direction is marked by the index *u*.

Table 7. Results of knit elongation in fine women's hosiery in the course direction – longitudinally

Yarn count	Loop sinking depth	ε_{eu}, %	$\Delta\varepsilon_{eu}$, %	ε_{pu}, %	$\Delta\varepsilon_{pu}$, %	$\Delta\varepsilon_{pu}$ - $\Delta\varepsilon_{eu}$ %	ε_{tu}, %	$\Delta\varepsilon_{tu}$ - $\Delta\varepsilon_{pu}$ %
20	**400**	180	56	230	71	15	323	29
20	**550**	240	71	250	74	3	338	26
20	**700**	230	72	240	75	3	319	25
20	**850**	270	73	290	79	5	369	21
33	**400**	200	67	220	74	7	298	26
33	**550**	170	63	200	75	11	268	25
33	**700**	200	66	220	73	7	302	27
33	**850**	170	57	220	73	17	300	27
40	**400**	150	62	170	70	8	243	30
40	**550**	140	55	160	63	8	255	37
40	**700**	150	56	190	71	15	266	29
40	**850**	170	64	190	71	7	267	29
60	**400**	170	57	200	67	10	300	33
60	**550**	160	58	190	69	11	277	31
60	**700**	150	58	180	69	12	260	31
60	**850**	150	60	180	73	12	248	27
			62±3		72±2	9±1		28±2

The main problem in measuring uniaxial tensile load of the knit is the very complex strain in the sample during tensile load. During elongation of a knit sample, the decrease of its width is uneven. At the point where it is fixed by grapples or clamps, the sample does not narrow. As the force and degree of elongation increase, the sample is becoming narrower and narrower in the middle. The point where the sample breaks is not always at the narrowest part. It often occurs at the point of contact with the grapple-clamp.

Peak loads or forces of breaking during strain are not important in the application of fine ladies' hosiery. Instead, the significant information is related to hosiery's elasticity and compressibility on the leg. When they put on a sock, women usually apply the force up to 60 N to pull the sock to a certain part of the leg. When the sock is pulled on the leg, the knit of the sock is unevenly formed, similarly to elongation on a dynamometer. Elongation of particular sections of a sock is regularly checked in the technological process of hosiery production. In order to measure hosiery elongation, manufacturers have internal regulations in sewing departments and a tabular display of extensibility of a particular section of hosiery under the force of 45 N. Some hosiery manufacturers have an elaborate system for measuring hosiery elongation at various loads. Under certain conditions they use tensile forces from 35 to 60 N with the 5 N step. For example, that is the method of measuring the length of elongation of smooth hosiery made of PA filament yarns in count 13 dtex under the tensile force of 35 N. Hosiery made of yarn in count 17 dtex is loaded with the tensile force of 40 N, or the hosiery made of yarns in count 60 dtex with the tensile force of 55 N. The tensile forces and elongations given are in some factories prescribed and adjusted to yarn counts and structures, parts of hosiery, knit structures, for unfinished, as well as finished or completed hosiery.

This paper gives the results of measuring structure parameters and elongation properties of knits after the knitting process. The knit is unfinished. In the process of production such knits are finished, i.e. they are subjected to chemical processes. After finishing, knitted fabrics change their properties, often around 3 %, and even up to 10 % for some of their properties. Therefore, pieces of hosiery that can be found on the market have somewhat different properties than the one discussed above.

7. Properties of completed hosiery

From the above mentioned it can be concluded that parameter values of knit structure in fine women's hosiery have a direct impact on hosiery elongation properties and thereby hosiery's functionality as well. Wearing comfort of hosiery is primarily reflected in the fit of a sock on the leg and its pressure on the leg.

Mass of material knitted in hosiery is often the main parameter for calculating hosiery production costs. A bigger mass knitted in hosiery will increase its selling price. Therefore it is very important to consolidate several parameters when designing hosiery in order to make good quality hosiery which will sell on the market at affordable prices. Bigger hosiery sizes require a bigger mass of knitted yarn making such hosiery more expensive on the market, cf. Table 8. For sock functionality it is necessary to consolidate fineness of yarn used for knitting, its elongation properties and the amount knitted into a certain size.

Table 8. Mass of fine ladies' pantyhose made in certain sizes and PA multifilament yarns of different fineness

Mass of pantyhose, g								
Size →		1 (36-38) XS	2 (38-40) S	3 (40-44) M	4 (44-46) L	5 (46-48) XL	6 (52-54) 2XL	7 (56-58) 3XL
Inlay →		without	small	small	small	large A	large B	saddle
Yarn count, dtex →	**22**	15.82	16.72	17.95	19.51	23.54	24.95	26.97
	33	22.83	24.06	24.72	26.19	31.29	32.69	34.82
	44	24.82	26.32	29.56	31.50	33.39	38.32	42.51
	60	43.10	45.02	49.39	54.68	59.84	61.27	66.48

8. Conclusion

Samples of fine ladies' hosiery were made using one automatic sock-knitting machine with four different loop sinking depths with the aim of determining elongation properties of hosiery. After the analysis of samples, the main conclusions are the following.

1. Increase of yarn count for hosiery production leads to the increase of mass per unit area of knitted fabric in hosiery (or hosiery mass) which is 38 to 99 g/m^2. Hosiery samples of mass per unit area of 34 to 38 g/m^2 were made of yarn in counts 20 dtex f 20 and samples of mass per unit area of 77 to 99 g/m^2 were made of yarn in counts 60 dtex f 60. This is important information because the price of hosiery depends on the amount of material used in its production.
2. Yarn count and loop sinking depth significantly affect the width of tubular hosiery shape which is 111 mm x 2 to 133 mm x 2.
3. Loop sinking depth has a significant influence on the parameters of yarn structure, primarily thread consumption per loop, which is at least 2.15 mm at the loop sinking depth of 400, and 3.21 mm at most at the loop sinking depth of 850.
4. Elongation properties of samples of fine women's hosiery produced depend on fineness of multifilament yarn they are made of, and primarily its elongation at break and the loop sinking depth at which hosiery is made. Elongation of the knit in the wale direction, or transverse elongation and elongation in the course direction, or elongation along hosiery were measured. During hosiery elongation three elongation areas were determined: elastic area, area of permanent deformation and boundary area connecting elastic and plastic area.

5. Elasticity of hosiery in the wale direction or transverse elasticity of hosiery is 60 to 260 % and depends on the yarn count and loop sinking depth at which the hosiery was made. Plastic area begins at hosiery elongation from 110 to 290 %. These two areas are significant during the use of hosiery. Elongation at break in the wale direction is very high and is 200 to 389 %. Elongation of hosiery in the course direction or elongation along hosiery does not have the same distribution as the elongation in the wale direction. Elasticity of hosiery in the course direction is 140 to 270 %. It slowly decreases with the increase of filament yarn count and is not so prominently dependent on loop sinking depth. Hosiery elongation in the course direction to the beginning of permanent deformation is 160 to 290 %, and elongation at break in the course direction 243 to 369 %. These are huge elongations that need to be adjusted in the construction of every sock.

The basic prerequisite in designing high-quality fine ladies' hosiery is to be familiar with elongation properties of yarns and structures of the knitted fabric in fine ladies' hosiery. Fineness and structure of filament PA yarn and loop sinking depth significantly influence the elongation properties of fine ladies' hosiery.

9. References

[1] Modig N 1988 *Hosiery Machines* (Meisenbach, Bamberg)
[2] Basnec I 1994 *Tekstil* **43** 255
[3] Mincev V 2006 *VGM1 - metod za ekspresno opredeljane na parametrite za pletene na čorapni izdelija* (Plovdiv, XLIII Congress of the IFKT)
[4] Lozo M Vrljičak Z 2014 MATRIB 2014 International conference on materials, tribology, recycling *Rastezna svojstva finih ženskih čarapa s gaćicama*, (Vela Luka) p 278-288
[5] http://www.strumpfhosen.at/content/5-groessentabelle
[6] http://www.lycra.com/ger_g/home.aspx
[7] Oess O 2014 2nd International Congress on Healthcare and Medical Textiles *Lycra Fiber T902C the Ideal Solution to Improve Comfort and Compliance of Medical Compression Products,* (Izmir)
[8] Cunko R and Andrassy M 2010 *Vlakna* (Zagreb, Tekstilno-tehnološki fakultet) p 240
[9] Koslovski A 2014 *Textil Lexikon*, (Stuttgart, Fachbuchverlag) p 220
[10] http://www.lonati.com/en-machines-item.asp?idf=2
[11] Vrljičak Z Kovač A 2011 *Tekstil* **60** 149
[12] Vrljičak Z Pavlović Z 2014 *Tekstil* **63** 27
[13] Kowalski K 2008 *Identifikacija procesu dziania na szydelkarkach* (Lodz, Polska Akademia Nauk, Oddzial w Lodzi, Komisja Wlokiennictwa)

IOP Conf. Series: Materials Science and Engineering **141** (2016) 012018 doi:10.1088/1757-899X/141/1/012018

Analysis of the heat setting process

N Besler[1], Y S Gloy[1] and T Gries[1]
[1] Lehrstuhl für Textilmaschinenbau und Institut für Textiltechnik (ITA) der RWTH Aachen University, Otto-Blumenthal-Str. 1, Aachen, Germany

E-mail: nora.besler@ita-rwth-aachen.de

Abstract. Heat setting is an expensive and energy elaborative textile process. Heat setting is necessary to guarantee size accuracy and dimensional stability for textile materials. Depending on the material different heat setting methods such as saturated steam or hot air are used for the fixation. The research aim is to define the influence of heat setting on mechanical characteristics and to analyse the correlation of heat setting parameters for polyester. With the help of a "one factor at a time" experimental design heat setting parameters are varied. Mechanical characteristics and the material quality of heat set and not heat set material are evaluated to analyse the heat setting influence. In the described experimental design up to a temperature of 195 °C and a dwell time of 30 seconds the material shrinkage of polyester is increasing with increasing temperature and dwell time. Shrinkage in wales direction is higher than in course direction. The tensile strength in course direction stays constant whereas the tensile strength in wales direction can be increased by heat setting.

1. Introduction

The Institut für Textiltechnik (ITA) der RWTH Aachen University is researching in the area of textile mechanical engineering, in textile simulation [1] as well as in quality control, automation [2] and energy efficiency [3]. The research fields in the area of knitting are improvements in pattern and process engineering [4]. ITA started researching on heat setting since July 2015 to analyse the process correlations. The heat setting process can be adjusted in a more efficient, innovative and user-friendly way based on fundamental process knowledge.

Heat setting is a thermal and mechanical process to guarantee size accuracy and dimensional stability for textile materials. Textile construction causes constraints on the textile. Constraints occur due to the arrangement of dipole links and hydrogen bonds, as well as crystallisation and chain stiffness. During the heat setting process these inner constraints are rearranged. Thus the dimension of the textile is changing. Depending on the material different heat setting methods such as saturated steam or hot air are used for the fixation. The standard method for the heat setting of synthetic material is the continuously used stenter frame. The stenter frame heat sets by using hot air. Important process parameters are temperature, dwell time, width and overfeed. For natural fibres relaxation dryers, crimping and steaming units are used [5-9].

Heat setting is an expensive and energy elaborative textile process. The correlation of heat setting parameters is analysed to determine the potential for process optimization. Models for optimum dimensional stability have been obtained for knitted cotton/elastane fabrics [10]. In this work further textile characteristics of polyester knitted samples are analysed with a "one factor at a time" experimental design. It is analysed how the textile characteristics (average weight, mesh density, thickness and tensile strength) change depending on the heat setting parameters. The relationship between temperature and time is always depending on the polymer and the textile characteristics such as fabric weight and construction [5].

2. Experimental setup

For the experimental design circular knitted polyester tube samples are used. The tubes are knitted on the machine BSM 3.0 from Beck GmbH, Albstadt. The machine has a diameter of 30" and a gauge of E24. The used polyester yarns have a specification of 167/ 48-1 CS. The tube samples have a size of 120 cm width and 60 cm height and have a single jersey structure. The average weight of the tube samples is 85 g/m² and the thickness is 0.5 mm. The mesh density in course direction is 12/cm and the mesh density in wales direction is 11/cm. The maximum tensile strength in course direction is 145 N and 214 N in wales direction.

A "one factor at a time" experimental design is carried out on the fixation press CC600 from Herbert Kannegiesser GmbH, Vlotho. The fixation press is working discontinuously. The polyester tube samples are placed on a belt. The samples are transported and heat set in course direction. The material is heated with heating panels from above and below. In addition the polyester samples can be pressed with the help of two pressure rolls. The pressure rolls are located behind the heating zone. After that the samples are falling onto a second belt, where the samples cool down. In the "one factor at a time" experimental design temperature and dwell time are varied on four levels, pressure is varied on two levels. The definition of the different levels is shown in Table 1.

Table 1. Level for heat setting parameters.

Level	Temperature [°C]	Dwell time [sec]	Speed [m/min]	Pressure [bar]
1	180	6.5	5.5	0
2	185	9	4	1
3	190	14.4	2.5	-
4	195	36	1	-

Textile characteristics are analyzed in order to see how heat setting parameters are correlating. The influence of heat setting on textile characteristics has to be identified to evaluate the correlation of heat setting parameters. Shrinkage is the most important quality factor for heat setting. Shrinkage is the dimensional change of textile occurring during heat setting. Analyzed textile characteristics are the average weight (DIN EN 12127), the mesh density (DIN EN 14971), and the thickness (DIN EN ISO 5048) to define the shrinkage. In addition the behavior of the tensile strength (DIN EN ISO 13834-2) is measured.

3. Results

During the knitting and the heat setting process many disturbance values occur. Disturbance values are for example variation in the knitting process, an imprecise temperature regulation in the fixation press, no possibility of material guidance during the heat setting process and thus occurring rolling of the edges. Due to these disturbance values high variation can be seen in the results and no assured statements can be made. Nevertheless tendencies of material behavior can be indicated.

The average weight is determined according to DIN EN 12127. The development of the average weight for the single jersey samples are shown in Figure 1. Five measurements are performed for each setting. The average weight of the reference material is 85 g/m². The average weight for heat set material is higher than for the reference material. Measurements are fluctuating around 87 g/m². During heat setting the material shrinks. Shrinkage means that the density of yarn increases. Thus the average weight is increasing as well. The average weight is increasing with temperature and time. The highest average weight of 90 g/m² is obtained at 195 °C and a dwell time of 36 seconds. Regarding additional pressure no assured influence on the average weight can be seen.

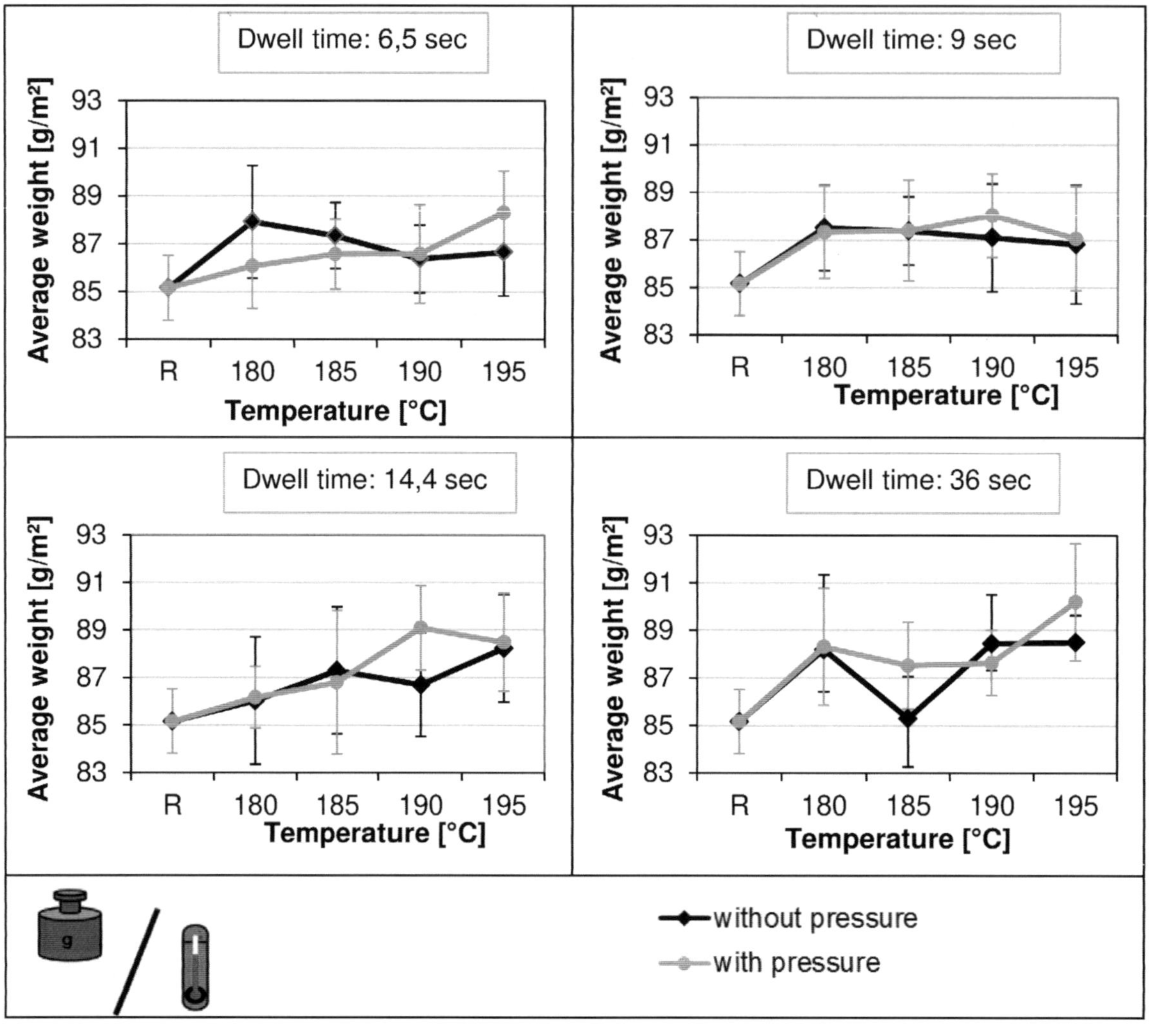

Figure 1. Average weight depending on temperature for different dwell times.

The thickness of the samples is determined according to DIN EN ISO 5048. Five measurements are performed for each setting. Figure 2 shows a decreasing thickness with increasing temperature. The same effect can be seen with increasing dwell time. Assuming that yarn is shrinking evenly, the size of the textile has to be reduced in all dimensions. Regarding the influence of pressure on the thickness no assured statements can be made. The standard deviation of the samples heat set with additional pressure is high. The course of the different curves indicates that pressure does not have an influence on the thickness of the material. A possible reason is the short time the material is exposed to the pressure rolls.

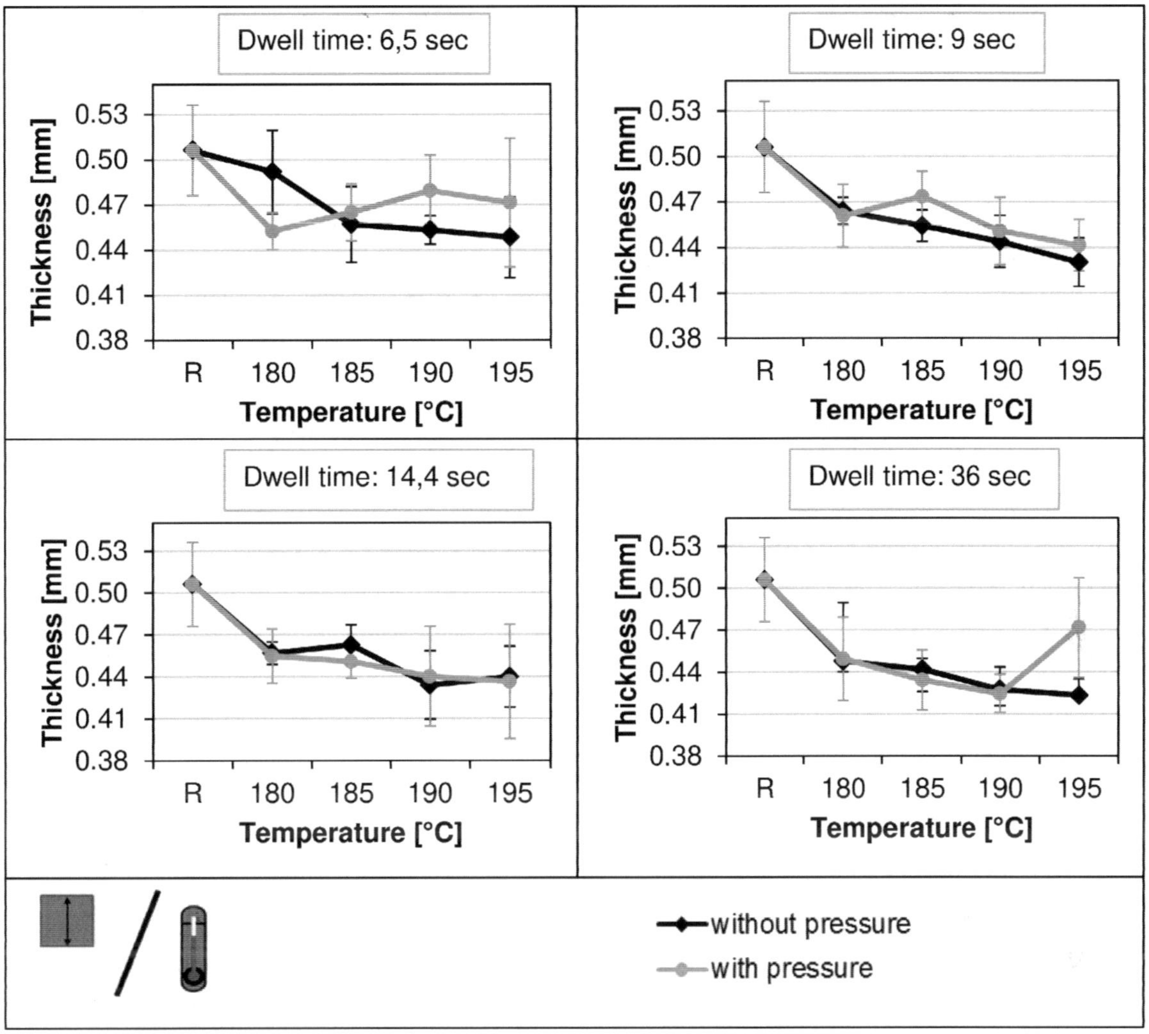

Figure 2. Thickness depending on temperature for different dwell times.

The mesh density is determined according to DIN EN 14971. Five measurements are performed for each setting. Figure 3 shows the mesh density in dependence of the temperature. The mesh density in course direction is constant (12/cm) for all temperature and dwell time levels. An influence of pressure is not visible. The mesh density in wales direction is increasing with increasing temperature and dwell time. The highest increase from 10.6/cm to 12/cm occurs at the highest temperature (195 °C) and the highest dwell time (36 sec) level. This means that the tested structures show a higher shrinkage in wales direction than in course direction. This can be explained by the unsymmetrical knitted mesh structure. The mesh structure is shown in Figure 4. The distance between two crossing points in wales direction is larger than in course direction thus the material in wales direction has more space to relax and shrink than the material in course direction. Another factor that needs to be considered is the reference mesh density. The mesh density in course direction is higher than in wales direction. On the one hand this means that there is more material having the possibility to shrink. On the other hand there is the possibility that the density is already so dense, that there is no possibility of further shrinkage due to a lack of space.

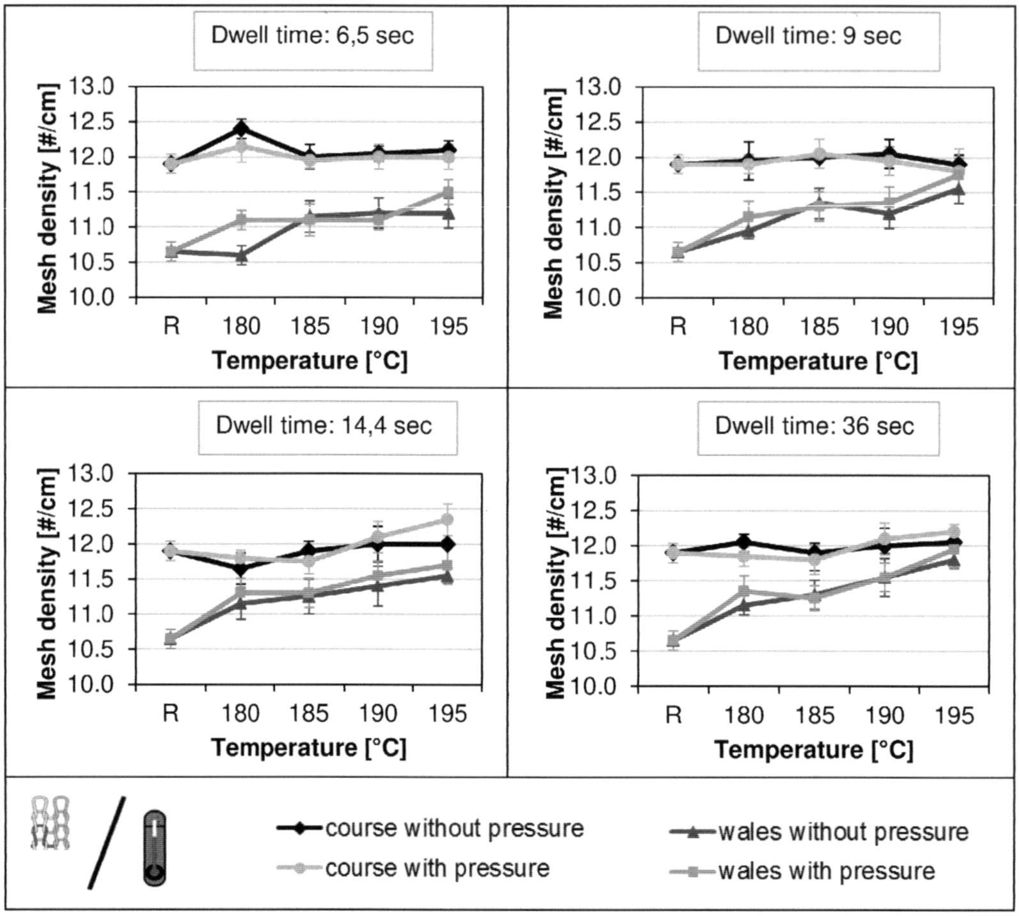

Figure 3. Mesh density depending on temperature for different dwell times.

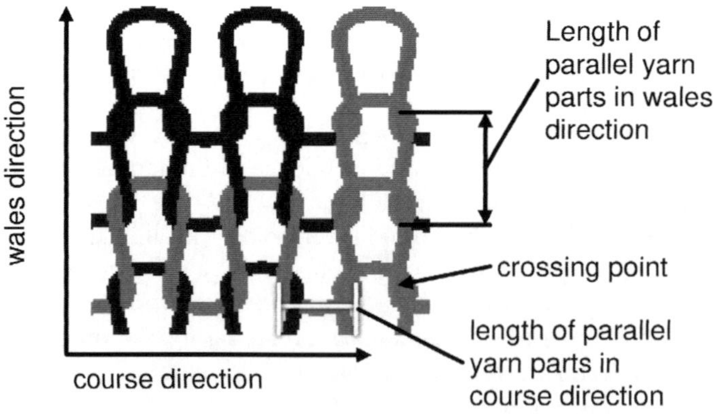

Figure 4. Knitted structure

Tensile strength is determined according to DIN EN ISO 13834-2. The results for the single jersey samples are shown in Figure 5. Five measurements are performed for each setting. For the single jersey samples anisotropic behavior of the tensile strength can be identified. The material behavior in course direction is different than in wales direction.

In course direction heat setting has a marginal influence on the tensile strength. The reference sample has a tensile strength of 145 N. The highest value reached for a heat set sample is 159 N, which is equivalent to a strength increase of about 10 %. The tensile strength is approximated to be constant for all heat setting level. An influence of pressure cannot be indicated. The tensile strength in wales direction is increasing with increasing temperature and dwell time compared to the reference sample. The maximum value for tensile strength in wales direction is 317 N, which is an increase of 48 % to the not heat set reference sample (214 N).

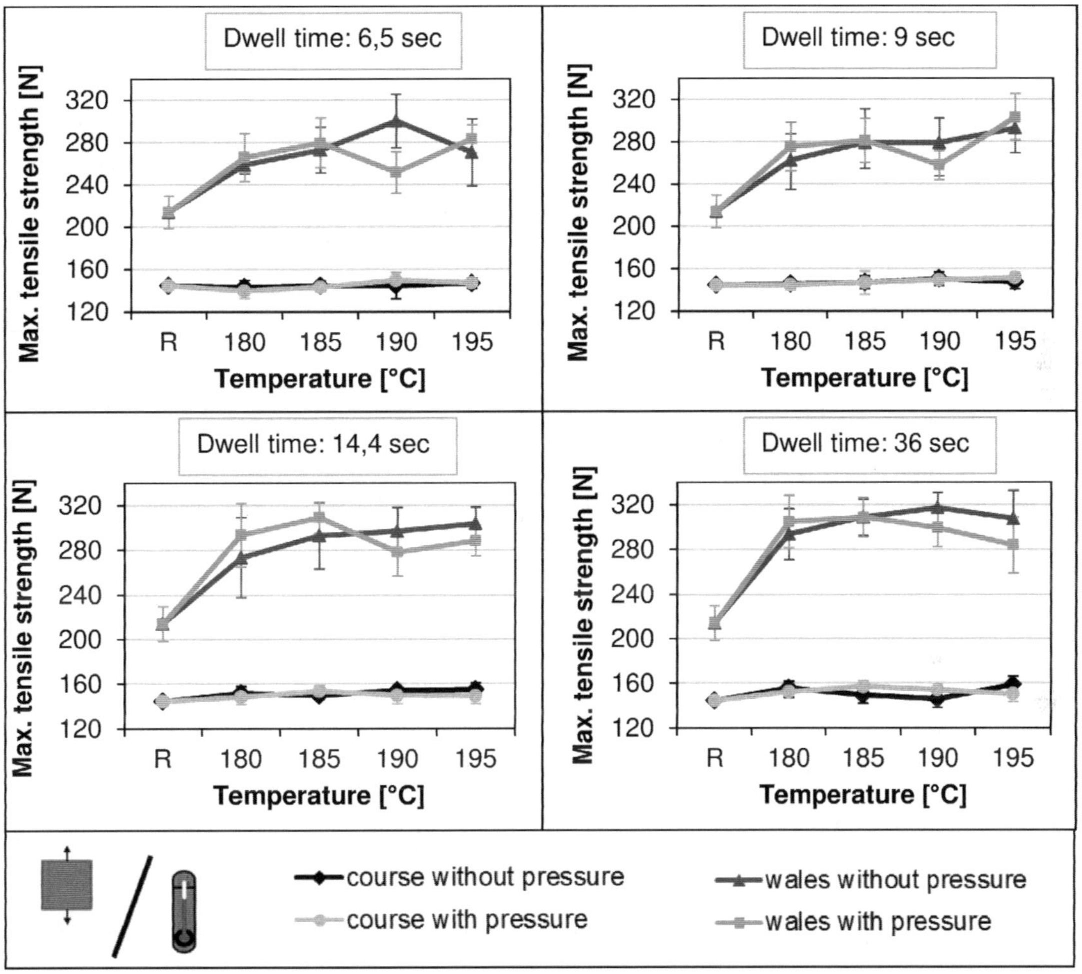

Figure 5. Tensile strength depending on temperature for different dwell times.

The tensile strength in course direction (145 N) is low compared to the tensile strength in wales direction (317 N). This result is equivalent to the results of the mesh density. With increasing mesh density in wales direction, there is more material taking up the forces of the tensile load. The tensile force can be distributed evenly. Furthermore filaments are taking the most loads in filament direction. Most parts of the filaments are oriented in wales direction thus more forces can be taken up in this direction (see Figure 4). Another cause for the increase of the tensile strength in wales direction is the transformation from amorphous into crystalline structures, which have a positive effect on the tensile strength. Higher temperatures and longer dwell times lead to a higher degree of crystallinity. Crystal growth in wales direction is encouraged by the knitted structure. The high amount of parallel yarn in wales direction develops more crystal structures than the short parallel yarn parts in course direction. The maximum tensile strength in wales direction grows [5].

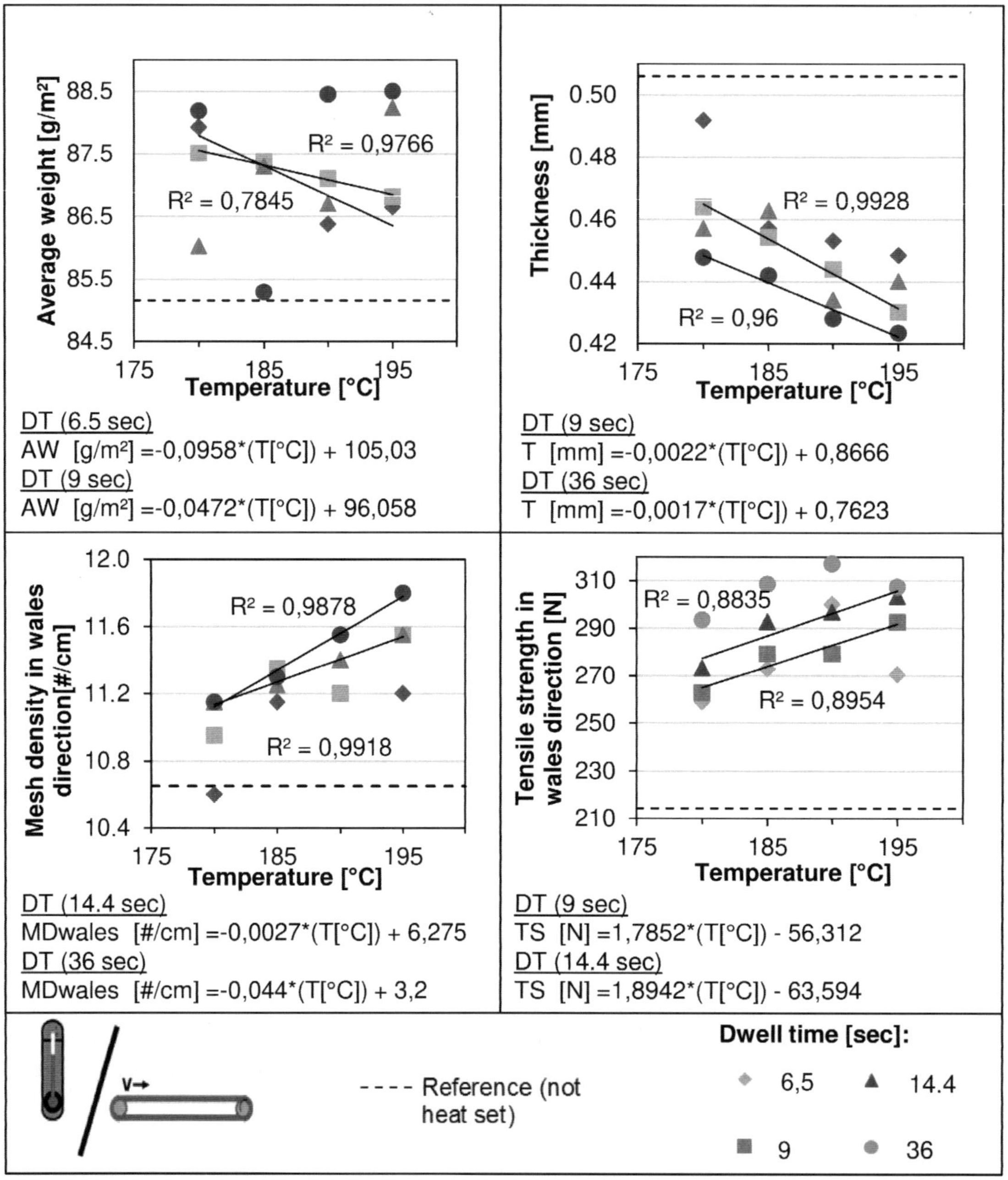

Figure 6. Correlation of dwell time and temperature for different characteristics.

A correlation between dwell time and temperature for each textile characteristic is shown in Figure 6. For the average weight linear regression for the dwell times of 6.5 sec and 9 sec and for the thickness for the dwell times of 9 sec and 36 sec can be approximated. For the mesh density (wales direction) the dwell times of 14.4 sec and 36 sec and for the tensile strength (wales direction) the dwell times of 9 sec and 14.4 sec can be approximated with linear regression. The linear regression models show that one desired textile characteristic (e.g. average weight) can be reached with different parameter combinations. Not all correlations can be approximated in linear regression models. More data have to be collected in order to define an appropriated model to describe the correlation between temperature and dwell time for each textile characteristic. The high standard deviations (see Figure 1-Figure 6) put the linear regression model into perspective.

4. Summary and Conclusion

In this paper the influence of heat setting on the textile characteristics of a polyester single jersey structure are analysed. The shrinkage is slightly increasing with increasing temperature (max. 195 °C) and increasing dwell time (max. 36 sec.). The mesh density and tensile strength in wales direction are increasing, whereas in course direction the increase is marginal and assumed to be constant. Pressure does not show any influence on the textile characteristics. To reach a desired textile characteristic different combinations of temperature and dwell time are possible.

For modelling the heat setting process further data has to be collected to improve and verify obtained models. For the engineering of technical textiles further analyses have to be done regarding the different behaviour of shrinkage and tensile strength in course direction and wales direction for different mesh densities. That way it can be analysed whether there is a mesh density limit for knitted structures. Another possibility is to analyse the crystalline structures to obtain more information about the increase of tensile strength. Furthermore results have to be transferred to the classical stenter frame heat setting and include washing tests to define the stable state.

Economic effects of the different heat setting parameters have to be analysed. Thus a data base with included economic effects can determine heat setting parameters for special textile characteristics. The data base can serve for the optimization of the heat setting process. The correlation of economic factors as well as quality factors can lead to conserving resources by energy reduction.

References

[1] Veit D 2012 *Simulation in Textile Technology 1st Edition* (Camebridge: Woodhead publishing limited)

[2] Saggiomo M et al 2014 *Integratives Messystem zur Fehlererkennung in der textile Flächenherstellung (OnLoom Imaging) in* Frey et al (Ed) *SPS IPC Drives: Elektrische Automatisierung; Systeme und Komponenten, Nürnberg 25.-27.11.2014* (Aachen: Primus)

[3] Holtermann T 2014 *Methode zur Bewertung und Erhöhung der Energieeffizienz von Produktionsprozessen der Textilindustrie* (Aachen: Shaker)

[4] Beer et al 2015 *Production technology selection for the development of technical fabrics in* Rimmer, Christine (Ed.) *Conference Proceedings / ATC-13 Asian Textile Conference 3-6 November 2015, Geelong, Victoria, Australia, Vol. 3, Part C. - Geelong: Deakin University, 2015, S. 1065-1068*

[5] Fontaine A 2011 *Technologie für Bekleidungsberufe* vol 14 (Köln: Bildungsverlag EINS GmbH) p 65

[6] Kaiser W 2011 *Kunststoffchemie für Ingenieure – Von der Synthese bis zur Anwendung* vol 3 (München: Carl Hanser Verlag) p 170

[7] Gries T et al 2014 *Textile Fertigungsverfahren – Eine Einführung* vol 2 (München: Carl Hanser Verlag) p 55

[8] Rouette H K 2001 *Encyclopedia of Textile Finishing* (Berlin Heidelberg: Springer)

[9] Horrocks A R and Anand S C 2000 *Handbook of Technical Textiles* (Cambridge: Woodhead publishing limited) p 161-164

[10] Nazir A et al. 2015 *Journal of Textile and Apparell, Technology and Management* Vol. 9 Issue 2 *Modelling Heat Setting of Cotton/Elastane Knitted Fabrics For Optimum Dimensional Stability*

Acknowledgments

We thank the Deutsches Zentrum für Luft- und Raumfahrt (DLR) and the Bundesministerium für Wirtschaft und Energie (BMWi) for funding this research project (funding no. 01MT15002C).

IOP Conf. Series: Materials Science and Engineering **141** (2016) 012019 doi:10.1088/1757-899X/141/1/012019

Effects of liquid ammonia treatment on the physical properties of knit fabric

I Y Lee[1,2], S D Kim[1] , C S Hwang[2], S R Kim[2] and S W Park[3]

[1] Department of Organic and Nano System Engineering, Konkuk University, Seoul, 143-701, Korea
[2] Korea High Tech Textile Research Institute, Yangju, 482-871, Korea
[3] Department of Advanced Fiber Engineering, Inha University, Incheon, 402-751, Korea

Abstract. The cellulosic knit fabric must be treated by NaOH solution in silket process to modify dyeability, luster, physical property, etc. But the silket treated knit fabric has a stiff touch, and must be treated with much of silicone softener. But it has bad durability of laundry. And the silket process has a problem that must discharge a lot of alkaline wastewater. In case of woven fabrics, as an alternative to silket process, liquid ammonia process was developed and this process is eco-friendly because the used ammonia is recovered by 98%. But the knit fabrics are not applicable to the conventional liquid ammonia process because they have selvedge curling problem and are very sensitive to tension. Recently, Korea High Tech Textile Research Institute(Korea) and Lafer SPA(Italy) worked together to develop the new liquid ammonia process for knit fabrics. In the present study, the physical properties of knit fabric after the newly-developed liquid ammonia treatment were investigated. The basic physical properties of knit fabric were measured using the Kawabata evaluation system. In addition, the dyeability, dimensional stability, eco-friendliness were investigated. The results showed that liquid ammonia treatment gave improved physical properties, which can be attributed to fast and uniform swelling, to knit fabric and resulted in a dimensional stability. The knit fabric treated in liquid ammonia showed a darker colour and unique appearance. Above all, the knit fabric treated in liquid ammonia had softer touch and superior gloss than the knit fabric of silket process. The new liquid ammonia process for knit fabrics will become the highest quality standard for knits and will be considered the preferred finish also thanks to eco-friendliness.

1. Introduction

Recently, a liquid ammonia process has been receiving much attention as a new environment friendly aftertreatment of woven fabric. The liquid ammonia has the boiling point of -33.4 degrees Celsius and the surface tension is 0.0334 N/m, which generates faster permeation than sodium hydroxide, reaching the swelling in a short period of time with uniform results after the treatment [1]. The reaction between liquid ammonia and cellulose has been known since 1930s but recently it has been receiving much attention in the aspect of excellent hand realization. In addition, its characteristics include low reduction of strength, high crease resistance, easy-care to cotton fabrics, and high recovery rate from the complete recycle system, making it an environment friendly process [2-4].

The cellulosic knit fabric must be treated by NaOH solution in silket process to modify dyeability, luster, physical property, etc. But the silket treated knit fabric has a stiff touch, and must be treated with much of silicone softener. But it has bad durability of laundry. And the silket process has a problem that must discharge a lot of alkaline wastewater.

In case of woven fabrics, as an alternative to silket process, liquid ammonia process was developed and this process is eco-friendly because the used ammonia is recovered by 98%. But the knit fabrics are not applicable to the conventional liquid ammonia process because they have selvedge curling problem and are very sensitive to tension.
Recently, Korea High Tech Textile Research Institute (KOTERI)(Korea) and Lafer SPA(Italy) worked together to develop the new liquid ammonia process for knit fabrics.
In the present study, the physical properties of knit fabric after the newly-developed liquid ammonia treatment were investigated. The basic physical properties of knit fabric were measured using the Kawabata evaluation system. In addition, the dyeability, dimensional stability, eco-friendliness were investigated.

2. Experimental

2.1. Materials
Cotton knit fabric used in the study is CM 30's single jersey knit with 60inch of width and 140g/m^2.

2.2. Conditions of liquid ammonia treatment of knit fabric
For the liquid ammonia treatment, the world-first commercial equipment for knit by Lafer spa, currently being operated in KOTERI, was used, and the ammonia was all evaporated by passing through the drying and steaming zone with the production rate of 18m/min. Drying and steaming temperature are 130 and 120 degrees Celsius, respectively. The drawing and actual photo of the world-first liquid ammonia equipment are as shown in Figure 1.

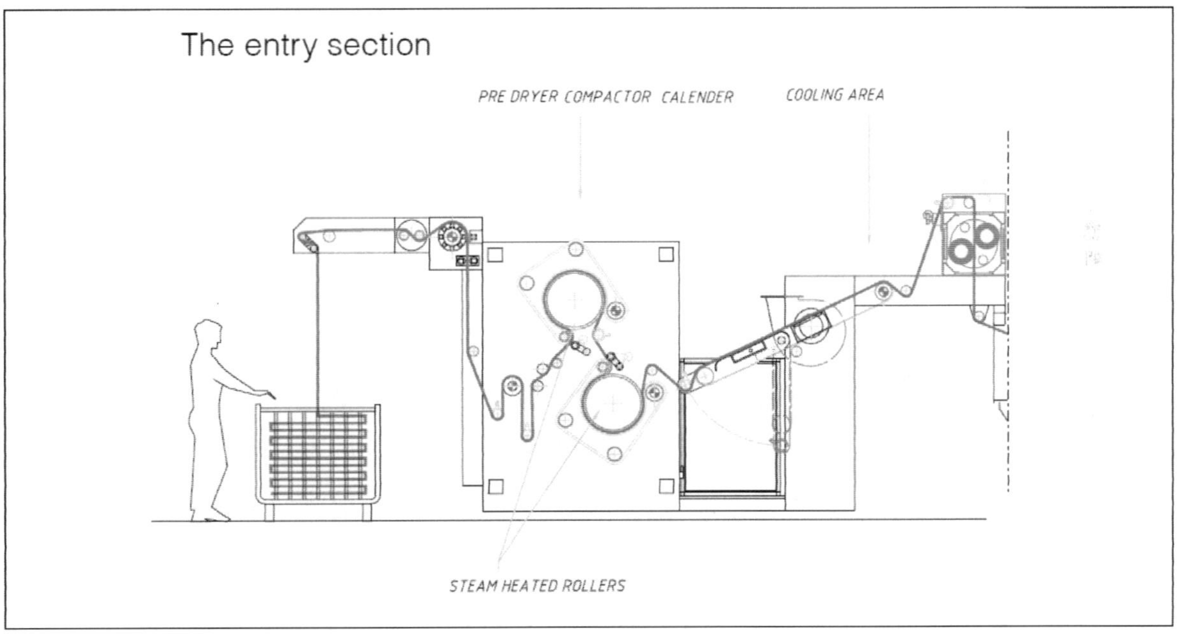

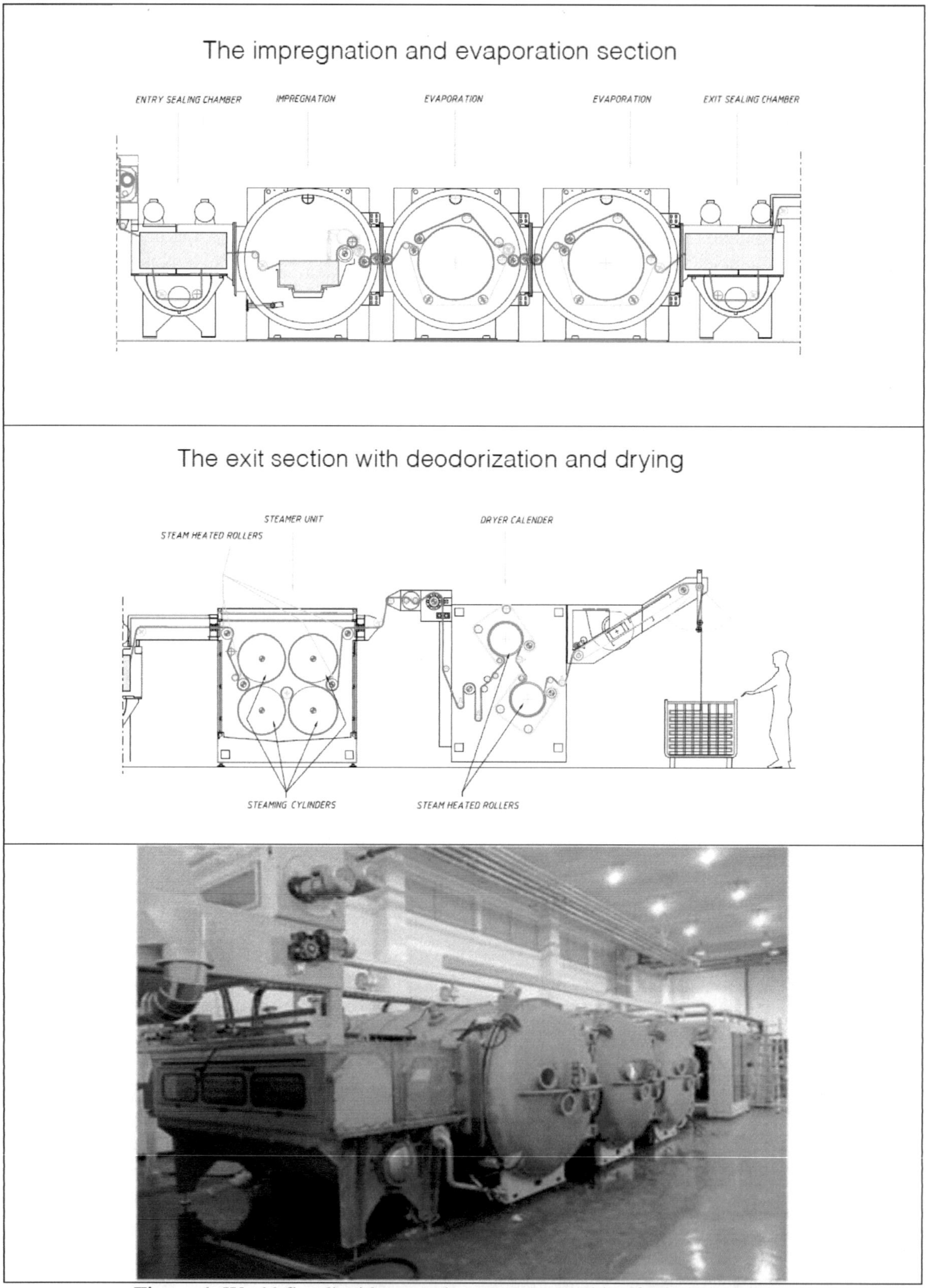

Figure 1. World-first liquid ammonia treatment machine for knitted fabrics

2.3. Dyeing conditions of knit fabric treated in liquid ammonia
For the dyeing of knit fabric, Cold-Pad-Batch(CPB) dyeing method was used. Dyeing conditions are as shown in Table 1 and the photos of the actual CPB process are presented in Figure 2.

Table 1. Dyeing conditions of knit fabric

Yarn type	Speed	Pick-up (%)	Batching Time	Batching Temp.
CM 30's	25m/min	100	16 Hours	25℃

Figure 2. CPB Dyeing process for knit fabric

2.4. Measurement of physical property
The measurement of physical properties was conducted using the KES-FB system (Kawabata Evaluation System for Fabric, Kato Tech. Co. LTD), which was suggested by the Hand Evaluation and Standardization Committee in Japan. The tensile property, bending property, shear property, compression property, surface property, and thickness and weight (6 properties, 17 items) were measured for each warp and weft directions at the standard conditions. The size of the samples used was 20×20cm, and samples were conditioned at standard condition prior to experiments. Study is concentrated on bending and shear properties that influence on the handle and touch of fabric. And bursting and tensile strength are measured separately.

2.5. Measurement of surface color
To measure the surface color changes by the process, the color difference meter (Color Eye 7000A, gretagmacbeth) was used. The spectral reflectance (R) was measured at the light source of D65, observation view of 10° and at the maximum absorbance wavelength, then K/S value was obtained using the Kubelk-Munk equation for evaluation of the dyeing property.

2.6. Measurement of dimensional stability
Dimensional stability was measured to evaluate the shrinkage of knit fabric after laundering. The test method is accomplished by AATCC 135.

2.7. Investigation of Eco-friendliness
Eco-friendliness was investigated to evaluate the manufacturing cost compared to the conventional process (mercerization).

3. Results and discussions

3.1. Physical property

3.1.1. Bending property

Bending property is an important property to determine various practical characteristics of fabric including the hand, drape, clothing formation performance, dimension stability, appearance of clothing and crease restoration. It is closely related to the fitness performance in the wearing process which is a continuous process of transformation and restoration, and a property to show the extent of fusion with the curve of human body. The bending property is expressed in bending strength (B) and bending hysteresis (2HB) [5]. The bending strength (B) is affected by contact pressure and density, contact field, thickness of thread between fibers and yarns. It has impact on the sewing as well as formation performance and hand of fabrics. A greater value means that the material is stiff and a smaller value represents outstanding performance of curve formation. Bending hysteresis (2HB) represents the proportional relations with the energy loss in the process of bending strains and restoration, where a smaller value means the extent of restoration from the bending change is more elastic. Results of the measurement are as shown in Figure 3. Untreated sample that means the pre-scoured knit fabric before liquid ammonia treatment showed the lower values indicating the liquid ammonia (LAMP) treated sample became a little stiffer after process treatment.

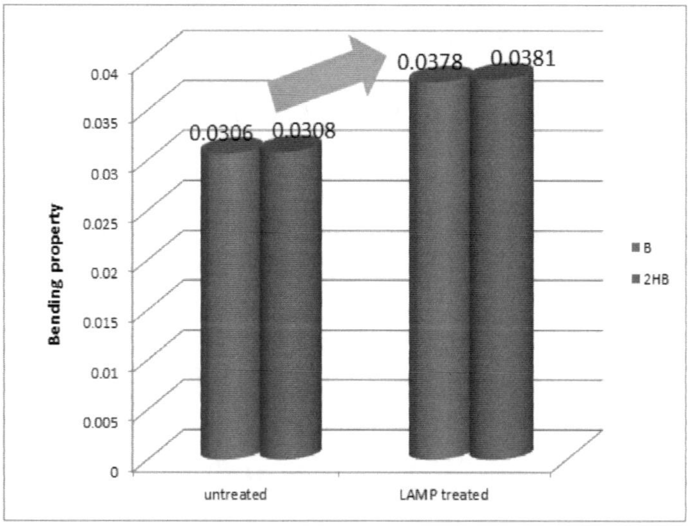

Figure 3. Results of bending property

3.1.2. Shear property

The shear property is a transformation to the external force that extended samples, and is a factor affecting the appearance of the curve or change in the human body when clothing is worn along with the bending property. Shear hysteresis is an important factor in the crease of fabrics [6]. Shear property is characterized by the shear stiffness (G), shear hysteresis (2HG). Shear stiffness (G) is a change by a shear force, and a small value means easy changes can occur. The shear hysteresis (2HG) and the hysteresis (2HG5) at the shear angle 5° represent the magnitude of energy loss for restoration after shear change. A small shear hysteresis means there is no difference in shear force at the initial direction and restored direction, showing elastic restoration. Results of the measurement are as shown in Figure 4. Untreated sample showed the lower values in broad lines indicating the liquid ammonia treated sample became a little stiffer and more elastic power after process treatment.

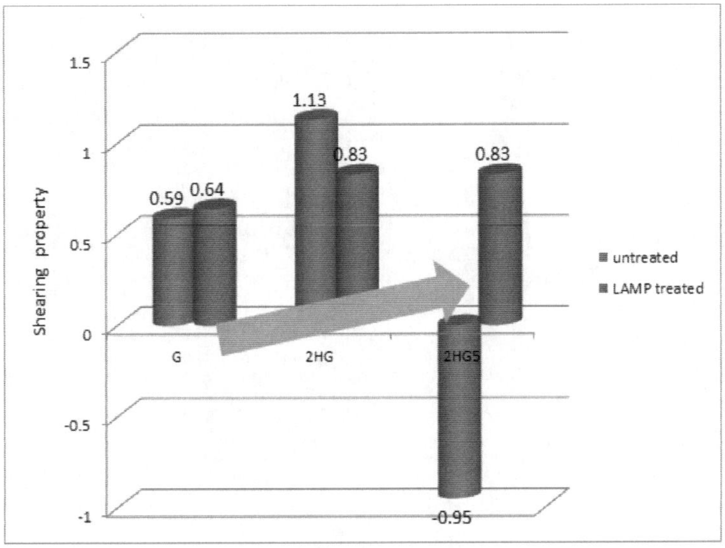

Figure 4. Results of shear property

3.1.3. Bursting and tensile strength

The bursting and tensile strength of fabric material depends on the linear density of thread, number of strands of wale and course. It is an important behavior to determine the drape, hand, sewing, wrinkles, and stability of fabrics [7]. Results of the measurement are as shown in Figure 5. Untreated sample showed the lower values indicating the liquid ammonia treated sample became stronger and more elastic power after process treatment.

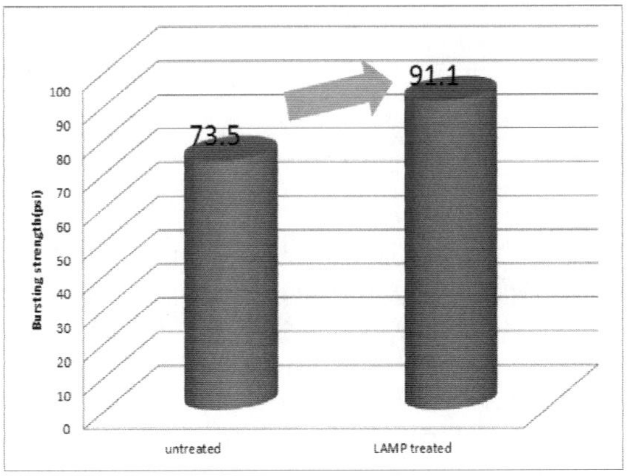

▲ Bursting strength

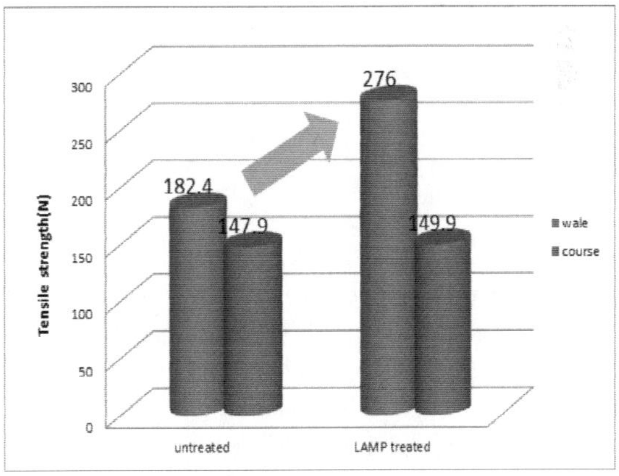

▲ Tensile strength

Figure 5. Results of bursting and tensile strength

3.2. Dyeability (Surface color)

Surface colors were measured to evaluate the changes of knit fabric colors by the process and the results are shown in K/S values. As shown in Figure 6, the LAMP treatment increased the K/S values showing darker colors.

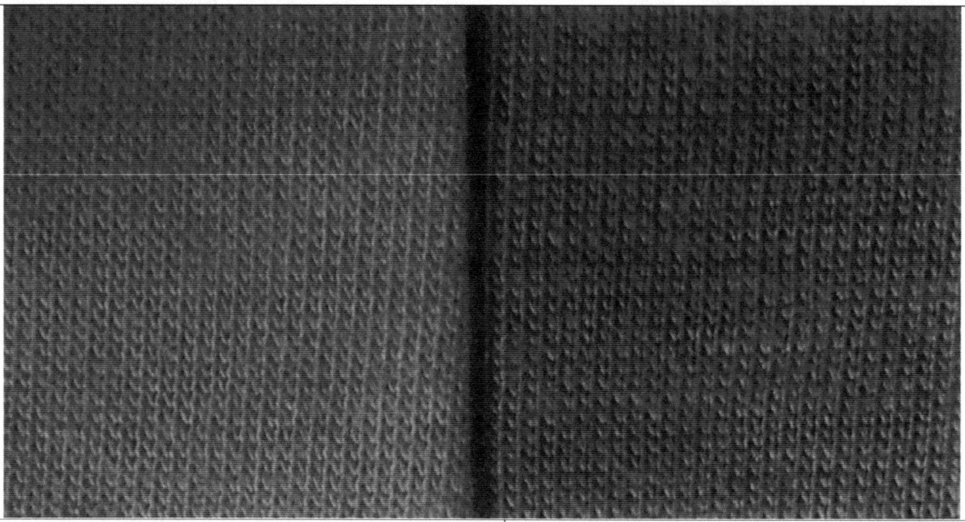

▲ untreated(K/S = 7.49)	▲ LAMP treated(K/S = 8.15)

Figure 6. Results of surface color change

3.3. Dimensional stability (Shrinkage)

Dimensional stability was measured to evaluate the shrinkage of knit fabric after laundering. As shown in Table 2, the LAMP treatment improved the shrink-proof property.

Table 2. Results of shrinkage

Shrinkge(%)		untreated	LAMP treated
Direction	Wale	-9.7	-1.7
	Course	-5.0	-4.1

3.4. Investigation of Eco-friendliness

Eco-friendliness was investigated to evaluate the manufacturing cost compared to the conventional process (mercerization). Save (%) was calculated with the cost for each row. As shown in Table 3, the LAMP treatment has a merit of eco-friendliness compare to the conventional process. In the view of total cost, while the total Mercerizing process cost is 309.7$, the total Liquid ammonia process cost is 73.82$. So the total save (%) is 76.2%.

Table 3. Results of Eco-friendliness and cost save(%)

	Mercerizing(Silket) process		Liquid ammonia process		Save(%)
	Demand amount	Cost($)	Demand amount	Cost($)	
Water	30.6ton	7.6	0.25ton	0.06	99.2
Steam	3.6ton	100	1.5ton	33.3	66.7
Auxiliary	450kg(NaOH)	171.9	0.7kg(NH3)	0.9	94.9
Electricity	180kWh	12.4	570kWh	39.4	-216.7
Wastewater	30.6ton	17.8	0.25ton	0.16	99.2
Total	-	309.7	-	73.82	76.2

IOP Conf. Series: Materials Science and Engineering **141** (2016) 012019 doi:10.1088/1757-899X/141/1/012019

4. Conclusion

In the present study, the physical properties of knit fabric after the newly-developed liquid ammonia treatment were investigated. The basic physical properties of knit fabric were measured using the Kawabata evaluation system. In addition, the dyeability, dimensional stability, eco-friendliness were investigated.

The results showed that liquid ammonia treatment gave improved physical properties, which can be attributed to fast and uniform swelling, to knit fabric and resulted in a dimensional stability. The knit fabric treated in liquid ammonia showed a darker colour and unique appearance. Above all, the knit fabric treated in liquid ammonia had softer touch and superior gloss than the knit fabric of silket process.

The new liquid ammonia process for knit fabrics will become the highest quality standard for knits and will be considered the preferred finish also thanks to eco-friendliness.

5. References

[1] I Y Lee, G E Jeong, S R Kim, C Bengelsdorff and S D Kim 2015 *Color. Technol.* **131** 192
[2] T Wakida, K Kida, M Lee, S Bae, H Yoshioka and Y Yanai 2000 *Text. Res. J.* **70** 328
[3] M C Lee, T Wakida, T Tokukama, C Doi, Y J Lim and S K Jeon 2005 *Text. Res. J.* **75** 13
[4] P F Greenwood 1987 *J. S. D. C.* **103** 342
[5] S A H Ravandi and M Ghane 2004 *J. Text. Inst.* **95** 277
[6] J Skelton 1976 *Text. Res. J.* **46** 862
[7] R C Dhingra, S D Jong and R Postle 1981 *Text. Res. J.* **51** 759

IOP Conf. Series: Materials Science and Engineering **141** (2016) 012020 doi:10.1088/1757-899X/141/1/012020

Design for Health and Well Being: Knitted Products for Diabetics

A Gault
Course Director Textile Art, Design and Fashion, York Street, Belfast, Northern, Ireland

E-mail: ae.gault@ulster.ac.uk

Abstract. This paper will discuss the design development, manufacturing and testing of knitted products maximizing the use of new innovations in Nano- technology and the integration of Phase Changing Materials specifically for diabetics. The project identified key aspects requiring design solutions to bring improvement to the circulatory problems with specific reference to the diabetic condition. Diabetics have particular difficulty in regulating their body temperature and this can result in the condition worsening, and resulting in loss of digits or limbs. The design of products to prevent the deterioration of the diabetic condition and to help those with limb loss was developed in collaboration with a Northern Ireland diabetic consultant, a product engineer and a knitwear designer. The fusion of ideas between the stakeholders resulted in the development and manufacture of a range of products that have been successfully tested at the yarn and fabric development stage and have been proven to maintain body temperature by either cooling or warming and therefore bring improvement to health and well-being. Whilst the product has a performance element the design ideas created desirable products that not only provided solutions to the brief but also resulted in products that had further market applications.

1. Introduction

The main focus of the paper is to investigate the outcomes and results of a project with a product designer, a diabetic consultant and a knitwear designer. This is in relation to the innovative developments in yarn technologies with particular reference to nanotechnology and phase changing materials. The needs of diabetics are clearly understood and articulated by a medical practitioner with many years of diagnostics and treatment. This expertise combined with the solution based and problem solving product engineer and the creativity, trend and market analysis of the designer. This triangulation and synergy in the group and research approach have all collaborated to produce a range of products specifically developed for diabetics. The products have been tested and are now commercially available.

Due to the research cluster identified target group were diabetics however it was recognized that there were potential benefits for a wider group of potential recipients. The range and diversity of products developed have included a cushion for a wheelchair, mattress, mattress cover, socks and garments including wraps. The emphasis of this paper is to investigate the garment designs and to assess if the design is fit for purpose combining an aesthetic quality that is equivalent to the technology involved.

As early as 2000 Bellucci identified that "the rapid progress in the field of nanoscience has been increasingly driving the attention of the scientific community as well as society at large on the corresponding technological applications, which are the object of so-called nanotechnology. A strong interest in assessing the current state of the art of this fast growing field as well as stimulating research networking." [1].

The selection of the nanoscience technology to design yarns manufactured with phase changing materials captured the expertise in the team and created a platform for developing and product creation. These products had the potential to benefit and improve health and well-being. The project while investigating the importance of a 'medical' product wanted to create products that would also be 'desirable'.

The number of people diagnosed with diabetes in Europe is estimated at 60 million, and the IDF (International Diabetic Federation) estimates that 46.5 % of adults remain undiagnosed. According to a Who Report (Diabetes Care 27: 1047- 1053, 2004) the global prevalence of diabetes and the number of all ages with diabetes for the years 2000 and 2030, show that the numbers of people with diabetes is projected to rise from 171 million in 2000 to 366 million in 2030, the IDF estimates that worldwide diabetes will rise to 642 million in 2040. The prevalence of diabetes is higher in men, but there are more women with diabetes than men. The most important demographic change to diabetes prevalence across the world appears to be the increase in the proportion of people >65 years of age. The findings indicate that the 'diabetes epidemic' will continue even if levels of obesity remain constant and that given the increasing prevalence of obesity, it is likely that these figures provide an underestimate of future diabetes. These figures and estimates show that the diabetic epidemic will bring financial burden to the health care worldwide [2].

The chronic nature of diabetes is costly for the health care system with both direct and indirect costs. The chronic nature of the condition and the gravity of its complications have meant that the indirect costs, such as the loss of productivity, disability, sick leave, early retirement and premature death are often higher than the direct costs. Studies in Latin America and America show that the tendency is 5 times higher (New*life*Outlook). Diabetes complications require hospitalization most of the time, and 50% of people with diabetes suffer from at least 1 complication. Hospitalization represents the biggest proportion of direct costs. It implies that admission to hospital, laboratory analysis, and the work of medical staff, specific therapies and investments [4-5].

Also according to New*life*Outlook, when you live with diabetes, it is not uncommon to feel cold or numb, especially in your extremities. Your core body temperature is closely tied to your metabolism, and since diabetes wreaks havoc on your metabolic processes, this in turn creates sweating, shivering, shaking far more than is evident in those without diabetes. Recent research in the National Health Service UK has uncovered the link between insulin and temperature. Insulin seems to work as an internal thermostat, helping to raise the core temperature by triggering the burning of 'brown fat' cells. Since insulin heals up the body it is no surprise that many type 1 diabetics (who have a depleted supply) should warm your body rather than cool it. However it is not quite so straightforward. One theory behind the chill of Type 2 diabetes centres on how long the insulin has been elevated. Chronically high insulin could exhaust the circuits that trigger the fat burning process, so the heat is never created, Since the body still needs heat for important metabolic reactions to take place, it begins to store insulating fat instead of burning fat. Extra fat on your body interferes with blood circulation, and that can leave your arms and legs, fingers and toes feeling particularly cold [3].

This difficulty in temperature regulation creates a tingling or burning pain that spreads from the fingers and toes up through the limbs. This can cause numbness, which can in turn lead to ulceration of the feet. Damage caused to the peripheral nervous system is termed, peripheral neuropathy. If the foot was to suffer any abrasions or cuts prevention of further damage or infection is of the utmost importance, as numbness from nerve damage can potentially cause one of the complications of diabetes that can result in limb amputation if the infection is not controlled [4].

It is this aspect of diabetes that the textile intervention seeks to reduce the possibility of either rubbing of the skin that might cause damage and further complications, and help regulate the body temperature through the investigation, development, design and manufacture of a series of products to help diabetics.

2. Experiments and methodology

As identified the target market were over 65, diabetic and while the products related to male and female the garments or wraps were tested with female diabetics. Market research, through qualitative and quantitative questionnaires was carried out with 100 females in this demographic; the same approach could be taken when assessing the products for mainly male sector and those products that are unisex. The focus of this research was to consider the level of satisfaction relating to the, weight, handle, comfort, maneuverability, colour and warmth.

2.1 Yarn

The yarn used to make the fabric was 5 ends of yarn (resulting 2/30's nm), this was polymers and copolymers with nanoparticles dispersed in the polymer matrix combined with 2 ends of Botanic fibres were integrated, cellulose yarn with the additive paraffin is refined plant oil. The reactive yarn technology created a micro heat accumulator with extraordinary latent heat potential. The phase changing material used is a micro composite with thermo regulating features, via direct spinning paraffin is embedded in a highly crystalline and tear resistant function this was combined with botanic fibres and integrated to use cellulose yarn with the additive paraffin is refined plant oil. [9]
Using the above series of processes to create yarns, they in turn were used to create a range of textile knitted garments including socks, wraps and seat covers. It is proposed based on previous research in the development of the yarns that textiles manufactured from these yarns can absorb excessive body-heat and when required, dispenses it again. Thus the temperature regulation is assured, providing extraordinary wearing comfort and excellent skin surface microclimate.

2.2 Structures

The fabrics were created on two different knitting machines, both weft knitting. The socks were produced on a Santoni sock machine (circular) and the wraps were produced on Shima Seiki flatbed machines. The garments used the polymer and copolymer yarns, combined with the cellulose yarns and were knitted on 12 gauge machines. The following structures were used to create the garments; all the yarn was kept the same throughout.

2 colour jacquard – stripe backing
3 colour jacquard – birds-eye backing 2 colour
jacquard – reverse jacquard 3 colour jacquard –
reverse Jacquard
Tensions were adapted to the stitch structure to enable correct qualities of knitting outcome.

2.3 Colour

Colour selection was an important part of the project. It was not only important to use the intelligent yarns to maximize comfort and promote temperature regulation. The methods for selecting colours was trend led through the use of WGSN (Worth Global Style Network) trend forecasting website, Lenzing colour predictions and Trend tablet. Market knowledge and understanding of colour was important for the end product selections to be effective and desirable, the investigation into colour was trend analysis and how colour selections were made. Colour (Scully and Johnston, 2012) has become increasingly important in the development, marketing and sales of goods, this is especially true for the fashion business and the consumer has become accustomed to a wide range of inspirational colour choices in apparel. According to the International colour authority "Colour comes before style and price, and is the first factor to which the customer responds" and is therefore important in satisfying the not only the medical intervention but the desirability to select but also to continue to choose to wear the product and the emotional response of something that is not simply a medical solution to a diagnoses.

Questionnaires were given to the target group of 100 female diabetics in the > 65 age range. Questions covered the following aspects of the wraps:

- Comfort in terms of weight in garment design when wearing for 3 hours
- Comfort in terms of maintaining body temperature (when donning garment design for approx. 3 hours).
- Comfort in terms of satisfaction of colour garment design and preference identified Comfort in terms of sensory satisfaction in garment design and handle – preference identified.
- Comfort in terms of maneuverability of design (for approx 3 hours donning)
- The wraps due to the structure of the garments and draping of the garment designs did have a variance of weights, these were as depicted in Table 1.

Table 1: Weight of Wraps

Results	Wrap 1	Wrap 2	Wrap 3	Wrap 4	Wrap 5
Weight	317.5g	330.0g	516.5	315.2g	528.0g

3. Results and Discussions

The following wraps were those selected for review and testing within the target group

Figure 1. Wrap 1 (Red, cream and Grey) and detail, wrap in 3 colour jacquard with birds-eye backing.

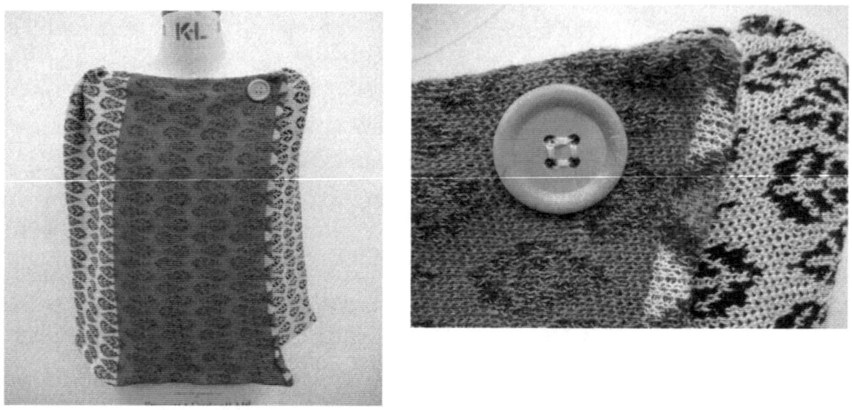

Figure 2. Wrap 2 (Blue, purple and cream) 2-colour jacquard birds-eye reverse

Figure 3. Wrap 3 (Grey, cream and black) 2-colour jacquard stripe reverse

Figure 4. Wrap 4 (Green, cream and black) 3 colour reverse (interlock) jersey structure

Figure 5. Wrap 5 (Multi-coloured) 3-coloured jacquard birds-eye reverse

Table 2. Quantitative Results of Questionnaires for preference and comfort selection. The numbers indicate the selected wrap identified via the questionnaires.

Wrap	Weight	Warmth	Colour	Handle	Movement
Wrap 1	22	19	17	17	19
Wrap 2	17	15	19	18	20
Wrap 3	10	12	21	18	12
Wrap 4	35	29	32	30	23
Wrap 5	16	25	11	17	26

The most popular wrap selected was Wrap 4 overall (in 4 elements) for weight, warmth, colour and handle but not for movement, that accolade went to wrap 5.

The results show that the preferred weight was the reverse interlock stitch, which has the highest weight out of all of the wraps due to the stitch structure. And was the most comfortable after continuously wearing for 3 hours. So the heaviest weight, the wrap containing the most yarn was the most warm. This structure also was the most pleasing in terms of handle, again due to the stitch structure the reverse interlock stitch. The most preferred wrap for movement was wrap 5 and this was the most comfortable to put on. The preferred colour was wrap 4, the green cream and black combination, all of the yarn content remained the same the difference came through stitch structure and style. The predicted colours in this target group were for purples according to trend information (WGSN, 2016) however the green wrap was the most on-trend colour in the selected range and the most desirable. Qualitative feedback is clear that colour was of utmost importance followed by the quality of handle and the warmth and comfort of the garment. The was an understanding that there were specific reasons to wear the garment, and that it would help temperature regulation, but it was clear by the comments that the majority would not wear a garment purely for functional value [7].

4. Conclusion

Therefore the 'diabetic epidemic' creates the necessity to reduce the impact of deterioration in the disease, to increase the diabetics' wellbeing but also to reduce direct and indirect costs. Interventions through the use of phase changing materials and the developments in nano technology can create yarns suitable to address those body-regulating issues that are prevalent in diabetics. However it is not enough to address this with medical textiles, there needs to be an element of market awareness and design, to provide this ever-growing group with a desirable product. The desirability of the product needs to meet all the requirements to ensure the success of the product regarding the key elements of comfort, weight, warmth, movement and colour. Further research needs to be developed for menswear and for other age demographics; this research was aligned to the UK market, with specific reference to the Northern Ireland region.

5. Selected Bibliography

[1] Bellucci S (ed) 2000 Proc. Of the School and Workshop on Nanotubes & Nanostructures (Santa Margherita di Pula (Cagliari), Italy, Sept. 2000)
[2] http://www.diabetesatlas.org/key-messages.html
[3] http://www.who.int/diabetes/facts/en/diabcare0504.pdf
[4] http://ec.europa.eu/health/major_chronic_diseases/diseases/diabetes/index_en.htm# fragment2
[5] http://ec.europa.eu/health/reports/docs/health_glance_2012_en.pdf
[6] http://dm2.newlifeoutlook.com/body-temperature-and-type-2-diabetes/
[7] http://www.wgsn.com/en/
[8] Scully K and Johnston Cobb D, Colour Forecasting for Fashion, 2012, Lawrence King, London.
[9] www.bioflexyarns.com

IOP Conf. Series: Materials Science and Engineering **141** (2016) 012021 doi:10.1088/1757-899X/141/1/012021

A novel medical bandage with enhanced clothing comfort

N Oğlakcioğlu, B Sari, T Bedez Üte, A Marmarali
Ege University, Department of Textile Engineering, Izmir, Turkey

Abstract. Compression garments are special textile products which apply a pressure on needed body zones for supporting medical, sport or casual activities. Medical bandages are a group of these garments and they have a very common usage for compression effect on legs or arms. These bandages are generally produced by using synthetic raw materials such as polyamide or polyester fibres. Medical bandages are in contact with skin. Even if the synthetic fibres are used, they may cause both comfort and health problems like allergies. Nowadays in textile sector, the expectations of clients include using of natural fibres as far as possible in all garments. Natural fibres have good advantages such as breathability, softness, moisture management ability, non-allergenic and ecologic structure and these characteristics present optimum utilization conditions. In this study, tubular medical bandages were manufactured by using core spun yarns (sheath fibres are selected as tencel, bamboo and cotton, core material is elastane) and their pressure and comfort (air and water vapour permeability) characteristics were investigated. The results indicated that the bandages have good comfort abilities beside adequate pressure values for compression effect. These garments can constitute a new production field for medical bandages with their comfort properties in addition to pressure characteristics.

1. Introduction

Compression therapy is a form of healing that has been utilized since the time of Hippocrates (450-350 BCE) and has been applied as therapy for treating diseases such as vein disorders or oedemas. The working mechanism of compression therapy is an external application which applies compression to the skin in order to support the superficial venous system. This pressure forces the veins to narrow and there is a reduction in the volume of blood in the veins. As a result, the calf muscle pump can work better and the bloodstream is more easily able to move up toward the heart. Thus there is higher tissue oxygenation and better micro circulation [1].

Compression garments have a special function, which apply a certain pressure to the body mainly for medical, sports and body shaping [2]. Most medical compression garments are individually designed and manufactured for a particular part of body, such as stockings, gloves, sleeves, face masks and body suits [3]. Medical bandages are strip or tube form of materials used to protect, immobilize, compress or support needed body parts and they are most common groups of compression garments. They provide a compression pressure by wrapping body parts to support an injured limb, muscle or joint. There are different types of compression bandages such as long stretch, short stretch etc. Medical bandages are cheap and common equipment for compression therapy but there are some difficulties of using for patients. Even if new bandage systems have easy utilization process, the users need an expert help to adjust right pressure values for a successful treatment [4-7].

The success rate of compression treatment is directly related to right pressure profile and permanently using [8]. Compression garments are in contact to skin very closely with stretch ability to support blood flow and muscle systems. This situation causes discomfort for patients particularly in summer condition. Because of direct contact, there is only a few microclimate between skin and textile materials, so new structures with advanced comfort properties are needed to be used for compression garments.

A comfortable textile product generally should not create excessive change on body temperature; efficiently remove the moisture and water vapour to the atmosphere and not cause irritation or allergies on skin [9]. Natural fibers are favoured raw materials for comfortable textile products with specialties such as breathability, high moisture absorbance, softness, non-allergenic and ecological structure. They are especially used in products which are in contact with skin just like underwear. Instead of synthetic fibers, using natural fibers increase day by day with the help of innovative trends in textile industry. New developed regenerated fibers provide the inspiration for outstanding products with their comfort and mechanical characteristics.

There are few studies about comfort characteristics of compression garments. Some of them examined the fiber characteristics of compression stocking and they indicated that using some different natural fiber or different fiber blend improved the comfort characteristics of stockings. In the study of Oğlakcıoğlu et al. investigated the effect of regenerated cellulose fibers on thermal comfort properties of compression stockings. The results showed that some of different special yarns such as viscose, modal or tencel could be recommended for summer stockings due to low thermal resistance, high water vapour and air permeability characteristics [10]. Bera et al. studied that the effect of varying the nylon and cotton blend percentage on comfort properties of pressure garments. According to results, the fiber blend percentage did not have any influence on air permeability and thermal property. However water vapour permeability and wicking behaviour affected significantly. Increasing in nylon percentage increased both of them [11].

A different study is about the air permeability, water transmission rate and thermal behaviour of pressure garments in extended condition to simulate the conditions during wear. This study revealed that the comfort properties changed significantly when fabric was held in an extended state. In this state the fabrics became thinner and this made the structure more permeable to air as well as water vapour [12]. Also in some other studies, the comfort characteristics of some special pressure garments using for hypertrophic burn scars were investigated. The new developed fabrics compared each other fabrics for comfort characteristics. Yıldız studied to generate a novel technique to determine pressure garments for hypertrophic burn scars and comfort properties. The thermo-physiological test results of this new composite fabric revealed that it had extremely low water vapour permeability and high resistance to evaporative heat loss. The water absorption percentage and absorption capacity were too low when compared control fabric which is a sportwool and single-jersey knitted [13]. Anand et al. investigated that the determination of the suitability of four fabrics to be utilized for management of hypertrophic scarring. The dimensional, mechanical and thermo-physiological properties of fabrics procured from different sources were compared. All fabrics also showed excellent water vapour permeability and extremely low resistance to evaporative heat loss from the skin to the environment [14].

Medical bandages are usually produced using polyamide covered elastane yarn as inlay yarn and polyamide filament yarn as ground yarn. Polyamide fiber is a suitable raw material with better mechanical properties and stretch ability but there are some disadvantages of this fiber such as soft touch and allergenicity. This study aims to improve comfort characteristics of medical compression bandages in addition to ecological effect, soft handle, anti-allergic features. In the study, three elastane core yarns were produced using cotton, bamboo and tencel fibers as sheath and tubular medical bandages were knitted with these elastane core yarns as single jersey structure for wrist zone. Produced bandages were compared with commercially available polyamide tubular bandages in order to determine their sufficiency about pressure characteristics. Also the thermal resistance, water vapour and air permeability of these knitted fabrics were statistically analyzed.

2. Material and method

In this study, initially three types of core-spun yarns were produced with count of 370 dtex. In these yarns 200 dtex elastane was used as core and three different types of cellulosic fibers were used as sheath. These fiber types were: cotton, bamboo and tencel. Yarns were produced on ring spinning frame equipped with core yarn apparatus, in pilot spinning mill of Ege University Textile Engineering Department. Yarn parameters are given in Table 1.

Table 1. Yarn production parameters

Raw Materials	Linear Density	T/m	Core Elastane Yarn Count	Elastane Feeding Tension
Cotton				
Tencel	370 dtex	550	200 dtex	3.5
Bamboo				

Then, tubular single jersey fabrics were knitted on a industrial stockinette machine with the same machine settings by these yarns. Table 2 illustrates the specifications of these fabrics and testing standards.

Table 2. Testing standards and fabric properties

Fabric Properties	Testing method	Material			
		Cotton	Bamboo	Tencel	Polyamide
Stitch density (stitch/cm^2)	EN 1049-2	240	246	234	120
Thickness (mm)	EN ISO 5084	1.6	1.7	1.7	1.8
Mass per area (g/m^2)	EN 12127:1997	475	470	440	595

Pressure characteristics were also determined for wrist zone since it is the most important parameter for these types of bandages. Pressure tests were performed on a pressure measuring device improved for compression stockings [15]. At the last stage, air permeability, water vapour permeability and thermal resistance values of cotton, bamboo and tencel fabrics were measured according to the related standards on Textest FX3300, Permetest and Alambeta instruments, respectively. The results were comparatively evaluated by ANOVA tests using SPSS software in order to determine thermal comfort levels of samples (Table 3). Beside these evaluations, the properties of developed fabrics were also compared to a commercially available polyamide tubular bandage to have an idea about pressure and comfort sufficiency.

3. Results and discussion

The statistical results revealed that using different yarn type (cotton, bamboo and tencel) significantly changes thermal comfort parameters. On the other hand, pressure results proved that all fabrics were in the same pressure class (class II), since they were knitted on the same machine by the same production adjustments (as seen in Table 2) with the same yarn properties (Table 1). The pressure and thermal comfort values and statistical differences for each yarn type are given in Tables 3. In this table, the mean values are marked with the letters ‘a’, ‘b’ and ‘c’. Any levels marked by the same letter showed that there is not any significant difference between the fabric types.

Table 3. Test results of measurements

Fabric Properties	Material			
	Cotton	Bamboo	Tencel	Polyamide
Pressure value (mmHg)	25.49[a]	25.98[a]	24.87[a]	28.99
Air permeability ($l/m^2/s$)	28.93[a]	50.01[b]	83.19[c]	55.88
Water vapour permeability (%)	36.48[a]	36.76[a]	38.28[b]	34.23
Thermal resistance (W/m^2K)	0.0226[a]	0.0310[b]	0.0294[b]	0.0256

Test results indicated that air permeability increase in cotton, bamboo and tencel fabrics, respectively, and tencel samples had the highest permeability within all samples (Figure 1). This result can be explained by lower hairiness and more compact structure of tencel yarn, as given in Figure 2, which provides higher fabric porosity and lower resistance to air penetration.

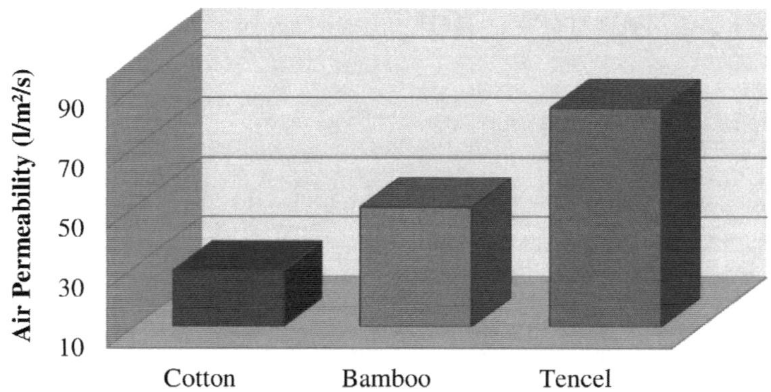

Figure 1. Air permeability results for different fiber types

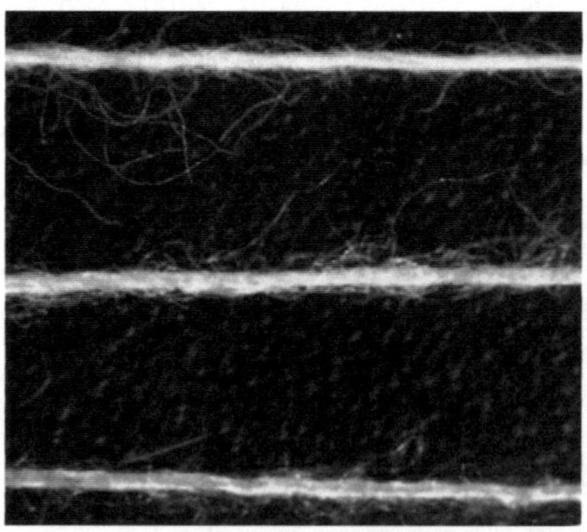

Figure 2. Microscopic yarn images

All samples have water vapour permeability values over 30%, which is accepted as limit for comfortable feeling. Whereas Tencel fabric achieves the highest water vapour permeability values as similar to air transfer property.

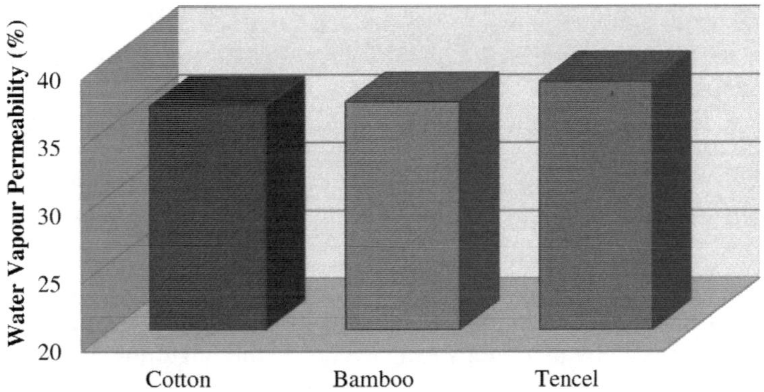

Figure 3. Water vapour permeability results for different fiber types

Cotton, bamboo and tencel are both cellulosic fibers, so it is expected them to have similar thermal conductivity values. However the entrapped air within the fabric structure significantly affects thermal properties [16]. The fabrics knitted by cotton yarns exhibit lower resistance to heat flow because of lower thickness and higher thermal conductivity. Since the thermal resistance property is proportional to the thickness of a layer and inversely proportional to its conductivity [17].

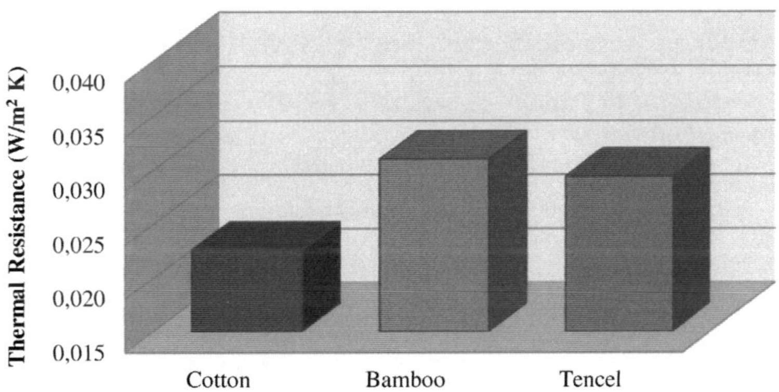

Figure 4. Thermal resistance results for different fiber types

4. Conclusion

Medical bandages are strip or tube form of materials used to protect, immobilize, compress or support needed body parts and they are most common members of compression garments. They are usually produced using polyamide covered elastane as inlay yarn and polyamide filament as ground yarn. In this study, elastane core yarn with cotton, bamboo and tencel fibers as sheath were produced and tubular medical bandage samples were knitted with these yarns.

The key factors of compression therapy are the right pressure profile and permanently usage. Compression garments apply higher pressure than a casual garment to the skin. Because of this feature, a small area for microclimate effect occurs between skin and fabric. This situation causes comfort problems for patient especially in warm conditions such as hot climate or during a physical activity. This research presents a new perspective of using natural core-spun yarns for medical bandage as a novel technique. In previous studies, some special or natural fibers were integrated to the compression garments as ground yarn, but together with polyamide covered elastane yarns. So these products showed a limited effect for comfort, because these fibers were not directly in contact with skin. In the scope of this study, the comfort properties of compression bandages are increased significantly with using natural fibers as sheath of elastic core spun yarns. According to the comfort test results, the developed samples show comparable properties (Table 1) than classical synthetic bandages. It is stated that tencel fibers are the most ideal raw material in summer days with higher air and water vapour permeability.

With this study, adequate pressure values were reached for compression therapy. The pressure results indicated that all knitted samples are in the same pressure class with commercially available polyamide bandages. It is proved that the compression garments at same pressure levels can be manufactured with suitable machine adjustments and elastane yarn counts even if different fiber types are used.

For further researches various core yarn types can be investigated for different aims such as burn treatment, leg ulcers. Because this technique enables to produce special compression garments with using elastane core yarn covered by different special fibers as raw material.

5. References

[1] Van Geest A J, Franken C Pand Neumann H A 2003 Medical elastic compression stockings in the treatment of venous insufficiency *Current Problems in Dermatology* **31**, 98-107

[2] Dias T, Cooke W and Fernando A 2006 Pressure garment *Patent US7043329B2* USA

[3] Wang L, Felder M and Cai J 2011 Study of properties of medical compression fabrics. *Journal of Fiber Bioengineering & Informatics* **4**, 15-22

[4] Krimmel G 2009 The construction and classification of compression garments *International Lymphoedema Framework* Template for Practice: compression garments for the upper body, 2-5

[5] Clark M 2003 Compression bandages: principles and definitions *Position Document – Understanding compression therapy* pp 5-7

[6] Choucair M and Phillips TJ 1998 Compression therapy *American Society for Dermatologic Surgery Inc* **28**, 141-148

[7] Vicaretti M 2010 Compression therapy for venous disease *Austrian Prescriber* 3(6) pp 186-190

[8] Ramelet A A 2002 Compression therapy *American Society for Dermatologic Surgery Inc.* **281**, 6-10

[9] Marmaralı A, Kretzschmar S, Özdil N and Oğlakcıoğlu N 2006 Parameters that affect thermal comfort of garment *Tekstil ve Konfeksiyon* **4**, 241-246

[10] Oğlakcıoğlu N and Marmaralı A 2010 Effects of regenerated cellulose fibers on thermal comfort properties of compression stockings *The Journal of Textile Engineering* **77**, 7-12

[11] Bera M, Chattopadyay R and Gupta D 2014 The effect of fibre blend on comfort characteristics of elastic knitted fabrics used for pressure garments *Journal of The Institution of Engineers (India):Series E* **95**, 41-47

[12] Gupta D, Chattopadyay R and Bera M 2011 Comfort properties of pressure garments in extended state *Indian Journal of Fibre& Textile Research* **36**, 415-421

[13] Anand S C, Govarthanam K K and Gazioglu D 2013 A study of the modelling and characterisation of compression garments for hypertrophic scarring after burns. part 2: characterisation of compression garments *The Journal of The Textile Institute* **1004**, 668-674

[14] Yıldız N 2007 A novel technique to determine pressure in pressure garments for hypertrophic burn scars and comfort properties, *Burns* **33**, 59-64

[15] Oğlakcıoğlu N and Marmaralı A 2014 An approach for measuring pressure characteristics of medical compression stockings *AATCC Journal of Research* **1**, 20-27

[16] Hes L 1999 Optimisation of shirt fabrics' composition from the point of view of their appearance and thermal comfort *International Journal of Clothing Science and Technology* **11**, 105-115

[17] Bogusławska Baczek M and Hes L 2011 The effect of moisture on thermal resistance and water vapour permeability of nomex fabrics *Journal of Materials Science and Engineering* **1**, 358-366